Florian Ion Petrescu

NEW

AIRCRAFT

USA 2012

Scientific reviewer:

Dr. Veturia CHIROIU
Honorific member of
Technical Sciences Academy of Romania (ASTR)
PhD supervisor in Mechanical Engineering

Copyright

Title book: New Aircraft

Author book: Florian Ion Petrescu

ISBN 978-1-4700-3404-7

WELCOME

Stealth aircraft are aircraft that use stealth technology to avoid detection by employing a combination of features to interfere with radar as well as reduce visibility in the infrared, visual, audio, and radio frequency (RF) spectrum. Development of stealth technology likely began in Germany during World War II. Well-known modern examples of stealth aircraft include the United States' F-117 Nighthawk (1981–2008), the B-2 Spirit, the F-22 Raptor, and the F-35 Lightning II.

While no aircraft is totally invisible to radar, stealth aircraft prevent conventional radar from detecting or tracking the aircraft effectively, reducing the odds of a successful attack. Stealth is the combination of passive low observable (LO) features and active emitters such as Low Probability of Intercept Radars, radios and laser designators. These are usually combined with active defenses such as chaff, flares, and ECM. It is accomplished by using a complex design philosophy to reduce the ability of an opponent's sensors to detect, track, or attack the stealth aircraft. This philosophy also takes into account the heat, sound, and other emissions of the aircraft as these can also be used to locate it.

You are welcome to read the full book!
The authors.

The Coandă-1910, designed by Romanian inventor **Henri Coandă**, was the first full-size attempt at a jet aircraft (fig. 1). Built as a sesquiplane, it featured an experimental aircraft engine which Coandă called the "turbo-propulseur," a centrifugal compressor propulsion system with a multi-bladed rotary fan situated in a duct and driven by a conventional piston engine. The unconventional aircraft attracted attention at the Second International Aeronautical Exhibition in Paris in October 1910, being the only plane without propeller. Coandă used a similar mechanism to drive a snow sled, but did not develop it further for aircraft.

Fig. 1. *The Coandă-1910, designed by Romanian inventor* **Henri Coandă**, *was the first full-size attempt at a jet aircraft.*

Decades later, after the practical demonstration of motorjets and turbojets, Coandă asserted that his turbo-propulseur was the first motorjet engine complete with fuel combustion in the air stream. He also said that he had made a single brief flight in December 1910, crashing just after take-off, the aircraft destroyed by fire.

About the Coanda's Effect

An early description of this phenomenon (Coanda) was provided by Thomas Young in a lecture given to The Royal Society in 1800:

The lateral pressure which urges the flame of a candle towards the stream of air from a blowpipe is probably exactly similar to that pressure which eases the inflexion of a current of air near an obstacle. Mark the dimple which a slender stream of air makes on the surface of water. Bring a convex body into contact with the side of the stream and the place of the dimple will immediately show the current is deflected towards the body; and if the body be at liberty to move in every direction it will be urged towards the current.

A hundred years later, Henri Coandă identified an application of the effect during experiments with his Coandă-1910 aircraft which mounted an unusual engine designed by Coandă. The motor-driven turbine pushed hot air rearward, and Coanda noticed that the airflow was attracted to nearby surfaces.

He discussed this matter with leading aerodynamicist Theodore von Kármán who named it the Coandă effect. In 1934 Coandă obtained a patent in France for a "Method and apparatus for deviation of a fluid into another fluid". The effect was described as the "Deviation of a plan jet of a fluid that penetrates another fluid in the vicinity of a convex wall."

The Coanda effect is a result of entrainment of ambient fluid around the fluid jet. When a nearby wall does not allow the surrounding fluid to be pulled inwards towards the jet (i.e. to be entrained), the jet moves towards the wall instead. The fluid of the jet and the surrounding fluid should be essentially the same substance (a gas jet into a body of gas or a liquid jet into a body of liquid). In one application, a jet of air is blown over the upper surface of an airfoil, which can have a strong influence on the overall lift, especially at high angles of attack when the flow would otherwise separate (stall).

The Coanda effect has important applications in various high-lift devices on aircraft, where air moving over the wing can be "bent down" towards the ground using flaps and a jet sheet blowing over the curved surface of the top of the wing. The bending of the flow results in its acceleration and as a result of Bernoulli's principle pressure is decreased; aerodynamic lift is increased.

The flow from a high speed jet engine mounted in a pod over the wing produces enhanced lift by dramatically increasing the velocity gradient in the shear flow in the boundary layer. In this velocity gradient particles are blown away from the surface, thus lowering the pressure there. Closely following the work of Coanda on applications of his research, and in particular the work on his "Aerodina Lenticulară," (fig. 2), John Frost of Avro Canada also spent considerable time researching the effect, leading to a series of "inside out" hovercraft-like aircraft where the air exited in a ring around the outside of the aircraft and was directed by being "attached" to a flap-like ring.

Fig. 2. *"Coanda* Lenticular aerodyne".

This is as opposed to a traditional hovercraft design, in which the air is blown into a central area, the plenum, and directed down with the use of a fabric "skirt". Only one of Frost's designs was ever built, the Avrocar.

The VZ-9 AV Avrocar (often listed as VZ-9) was a Canadian vertical takeoff and landing (VTOL) aircraft developed by Avro Aircraft Ltd. as part of a secret U.S. military project carried out in the early years of the Cold War.

The Avrocar intended to exploit the Coandă effect to provide lift and thrust from a single "turborotor" blowing exhaust out the rim of the disk-shaped aircraft to provide anticipated VTOL-like performance. In the air, it would have resembled a flying saucer. Two prototypes were built as "proof-of-concept" test vehicles for a more advanced USAF fighter and also for a U.S. Army tactical combat aircraft requirement.

The effect was also implemented during the U.S. Air Force's AMST project. Several aircraft, notably the Boeing YC-14 (the first modern type to exploit the effect), have been built to take advantage of this effect, by mounting turbofans on the top of wing to provide high-speed air even at low flying speeds, but to date only one aircraft has gone into production using this system to a major degree, the Antonov An-72 'Coaler'.

The McDonnell Douglas YC-15 and its successor, the Boeing C-17 Globemaster III, also employ the effect. The NOTAR helicopter replaces the conventional propeller tail rotor with a Coanda effect tail (fig. 3).

Fig. 3. *The C-17 Globemaster III uses the Coanda effect for a comfortable ride at low flying speeds*

An important practical use of the Coanda effect is for inclined hydropower screens, which separate debris, fish, etc., otherwise in the input flow to the turbines. Due to the slope, the debris falls from the screens without mechanical clearing, and due to the wires of the screen optimizing the Coanda effect, the water flows though the screen to the penstocks leading the water to the turbines.

The Coanda effect is also used to make automotive windshield washers which function without moving parts and to create pneumatic logic circuits.

The operation principle of oscillatory flowmeters also relies on the Coanda phenomenon. The incoming liquid enters a chamber that contains 2 "islands". Due to the Coanda effect the main stream splits up and goes under one of the islands. This flow then feeds itself back into the main stream making it split up again, but in the direction of the second isle. This process repeats itself as long as the liquid circulates the chamber, resulting in a self induced oscillation that is directly proportional to the velocity of the liquid and consequently the volume of substance flowing through the meter. A sensor picks up the frequency of this oscillations and transforms it into an analog signal yielding volume passing through.

In air conditioning the Coanda effect is exploited to increase the throw of a ceiling mounted diffuser. Because the Coanda effect causes air discharged from the diffuser to "stick" to the ceiling, it travels farther before dropping for the same discharge velocity than it would if the diffuser was mounted in free air, without the neighbouring ceiling. Lower discharge velocity means lower noise levels and, in the case of variable air volume (VAV) air conditioning systems, permits greater turn-down ratios. Linear diffusers and slot diffusers that present a greater length of contact with the ceiling exhibit greater Coanda effect.

In meteorology, the Coanda effect theory has also been applied to some air streams flowing out of mountain ranges such as the Carpathian Mountains and Transylvanian Alps, where effects on agriculture and vegetation have been noted. It also appears to be an effect in the Rhone Valley in France and near Big Delta in Alaska.

Fig. 4. *The McDonnell Douglas YC-15 uses the Coanda effect for a comfortable ride at low flying speeds*

The YC-15 was McDonnell Douglas' entrant into the U.S. Air Force's Advanced Medium STOL Transport (AMST) competition (fig. 4), to replace the C-130 Hercules as the USAF's standard STOL tactical transport. In the end neither the YC-15 nor Boeing YC-14 was ordered into production, although the YC-15's basic design would be used to form the successful C-17 Globemaster III.

The feathered propellers of an RAF Hercules C.4. The NOTAR helicopter replaces the conventional propeller tail rotor with a Coanda effect tail (fig. 5).

Fig. 5. The NOTAR helicopter replaces the conventional propeller tail rotor with a Coanda effect tail.

Several aircraft, notably the Boeing YC-14 (the first modern type to exploit the effect), have been built to take advantage of this effect, by mounting turbofans on the top of wing to provide high-speed air even at low flying speeds (fig. 6).

Fig. 6. *The Boeing YC-14 use the Coanda effect.*

The Avro Canada VZ-9 Avrocar was a VTOL aircraft developed by Avro Aircraft Ltd. (Canada) as part of a secret U.S. military project carried out in the early years of the Cold War (fig. 7). The Avrocar intended to exploit the Coandă effect to provide lift and thrust from a single "turborotor" blowing exhaust out the rim of the disk-shaped aircraft to provide anticipated VTOL-like performance. In the air, it would have resembled a flying saucer. Two prototypes were built as "proof-of-concept" test vehicles for a more advanced USAF fighter and also for a U.S. Army tactical combat aircraft requirement. In flight testing, the Avrocar proved to have unresolved thrust and stability problems that limited it to a degraded, low-performance flight envelope; subsequently, the project was cancelled in September 1961.

Fig. 7. The Avro Canada VZ-9 Avrocar was a VTOL aircraft developed by Avro Aircraft Ltd. (Canada). The Avrocar intended to exploit the Coanda effect to provide lift and thrust from a single "turborotor" blowing exhaust out the rim of the disk-shaped aircraft to provide anticipated VTOL-like performance.

Through the history of the program, the project was referred to by a number of different names. Avro referred to the efforts as Project Y, with individual vehicles known as Spade and Omega.

Project Y-2 was later funded by the US Air Force, who referred to it as WS-606A, Project 1794 and Project Silver Bug. When the US Army joined the efforts it took on its final name "Avrocar", and the designation "VZ-9", part of the US Army's VTOL projects in the VZ series.

The Avrocar was the ultimate result of a series of blue skies research projects by designer "Jack" Frost, who had joined Avro Canada in June 1947 after working for several British firms. He had been with de Havilland from 1942 and had worked on the de Havilland Hornet, de Havilland Vampire jet fighter and the de Havilland Swallow aircraft, where he had been the chief designer on the supersonic research project. At Avro Canada, he had worked on the Avro CF-100 before creating a research team known as the "Special Projects Group" (more commonly known as SPG). Frost first surrounded himself with a collection of like-minded "maverick" engineers, then arranged for a work site. Initially ensconced in the "Penthouse" (the derisive company nickname for the executive wing) of the Administration Building, the SPG was subsequently relocated to a Second World War-era structure across from the company headquarters, the Schaeffer Building, that was secured with security guards, locked doors and special pass cards. At times, the SPG also operated out of the Experimental Hangar where it shared space with other esoteric Avro project teams.

At the time, Frost was particularly interested in jet engine design and ways to improve the efficiency of the compressor without sacrificing the simplicity of the turbine engine. He found Frank Whittle's "reverse flow" design too complex and was interested in ways to "clean up" the layout. This led him to design a new type of engine layout with the flame cans lying directly outside the outer rim of the centrifugal compressor, pointed outwards like the spokes on a wheel. Power for the compressor was drawn from a new type of turbine similar to a centrifugal fan, as opposed to the more typical propeller-like turbine, driving the compressor using gearing rather than a shaft. The resulting engine had no conventional thrust axis, and was arranged in the form of a large disk, which he referred to as a "pancake engine." The jet thrust exited from around the entire rim of the engine, and this presented problems trying to adapt the design to a typical aircraft.

At the same time, the aircraft industry as a whole was becoming increasingly interested in VTOL aircraft. It was expected that any future European war would start with a nuclear exchange that would destroy most airbases, so aircraft would need to operate from limited airbases, roads or even unprepared fields. Considerable research effort was put into various solutions to securing a second-strike capability. Some of these solutions included rocket-launched aircraft like the zero-length launch concept, while many companies started work on VTOL aircraft as a more appropriate long-term solution.

Frost felt the excellent performance of his new engine would be a natural fit for a VTOL aircraft due to its high expected power-to-weight ratio. The problem was how to use the annular thrust to drive the aircraft forward, as well as the problem of fitting the very large engine into a suitable airframe. Frost suggested using a series of vents to redirect the thrust flowing out of the "front" of the engine towards the rear, although it was well known that long channeling leads to a loss of thrust. In order to keep the "piping" as short as possible, the design ported the thrust out along the leading edge of what was essentially a very large delta wing. As the engine was disk-shaped, the triangular shape was "pushed out" near the front, producing a planform shaped roughly like a spade. For this reason the design was also referred to as the "Avro Ace," a likely reference to the Ace of Spades. The compressor inlet was located at the middle of the engine, so the engine air intakes were located just to the front of the centre on the top and bottom of the aircraft. The cockpit was positioned over the main bearing, behind the intakes. A "spine" on the top and bottom ran from the cockpit area to the rear edge of the aircraft. Several other versions of the basic layout were also studied, including the "Omega" which was more disk-like as it cut away the rear portions of the delta wing as well.

For VTOL operations the aircraft was expected to sit pointed up, supported by long landing legs that extended out of the spine. Landing would be accomplished at a very high angle, making visibility during the approach very difficult. A number of other VTOL experiments of the era attempted various solutions to this problem, including rotating pilots seats and cockpits, but none proved very effective. Another problem with various VTOL experiments was that stability in a hover was difficult to arrange, although not entirely unexpected. A solution to this problem would require the thrust to be directed downward from a larger area, as it is in a helicopter, where the lift is supplied over the entire area of the rotor disk. Most designers turned to bleeding off air from the engine's compressor, and directing that through pipes arranged around the aircraft. Frost's engine design used such a large number of nozzles that such an arrangement would not be to easy to build.

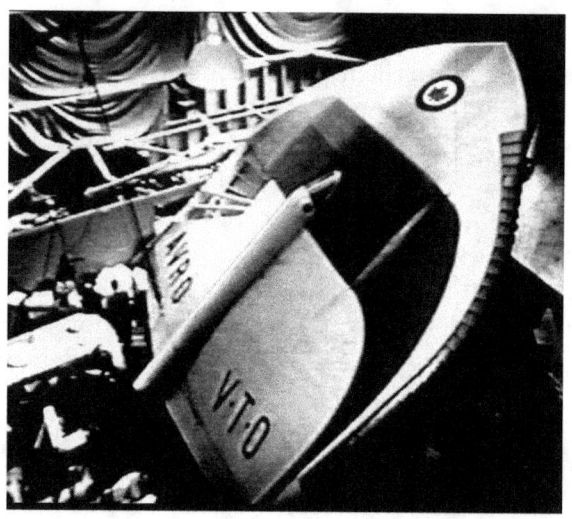

Fig. 8. *VTOL aircraf project Y was capable to fly with 1,500 miles per hour (2,400 km/h) and climbing vertically.*

In 1952, the design was advanced enough that the Canadian Defense Research Board funded the effort with a $400,000 contract. By 1953, a wooden mock-up of Project Y was completed, of which only images remain (fig. 8). It appears the project was considered too costly within the military establishment, which was at the time involved in several extremely expensive air defense projects. On 11 February 1953, a story on the project was leaked to the Toronto Star along with images of the Omega design, apparently in order to gain further funding (a strategy widely employed in the U.S. at the time, known as policy by press release). Five days later, the Minister for Defense Production informed the House of Commons that Avro was indeed working on a "mock-up model" of a flying saucer, capable of flying at 1,500 miles per hour (2,400 km/h) and climbing vertically. Nevertheless, further funding was not forthcoming.

While Project Y continued, Frost had meanwhile become interested in the Coandă effect, where fluid flows will follow strongly convex shapes, something that might be unexpected at first glance. Frost felt the effect could be used with his engine design to produce a more practical VTOL aircraft, the exhaust flowing outward over the upper surface of the aircraft and then being directed downward over a flap-like arrangement. This would produce a lift force around the entire edge of the aircraft, allowing it to land "flat". He produced a number of small experimental designs using compressed air in place of an engine in order to select a suitable planform shape, and eventually decided that a disk was the best solution.

As he continued these experiments, he found that the same thrust-direction system he intended for VTOL operations worked just as well for forward flight. In this case the disk shape was not of itself a good lifting surface, as it was neutral in terms of lift direction — that is, it would fly sideways as readily as it would fly forward. However, by modifying the airflow with the application of a small amount of jet thrust, the overall airflow over the craft could be dramatically altered, creating a sort of "virtual airfoil" of any needed configuration. For instance, by directing even a small amount of jet thrust down, a large mass of air would be pulled over the upper surface of the wing and dramatically augment the flow over the wing, creating lift.

This appeared to offer a solution to one of the most vexing problems of the era, designing an aircraft that was effective at subsonic and supersonic speeds. Subsonic lift is created by the airflow around the wing following streamlines, but supersonic lift is generated by shock waves at points of critical curvature. No single design could offer high performance for both regimes. The blown disk could attack this problem by being laid out for supersonic performance only, and then using jet thrust to modify subsonic airflow into a semblance of a normal wing. The resulting design would be tuned for high supersonic performance, have reasonable subsonic performance, and would also offer VTOL, all in a single design.

In late 1953, a group of U.S. defence experts visited Avro Canada to view the new CF-100 fighter jet. Somewhere along the way, Frost co-opted the tour and rerouted it to the Special Projects area where he proceeded to show off the Project Y mock-up and models and drawings (some never before seen by senior company officials) for a completely circular disk-shaped aircraft known as "Project Y-2." The USAF agreed to take over funding for Frost's Special Projects Group, and a contract for $US 750,000 followed in 1955. By 1956, Avro management was interested enough to commit $2.5 million to build a "private venture" prototype. In March 1957, the Air Force added additional funding, and the aircraft became Weapons System 606A (fig. 9).

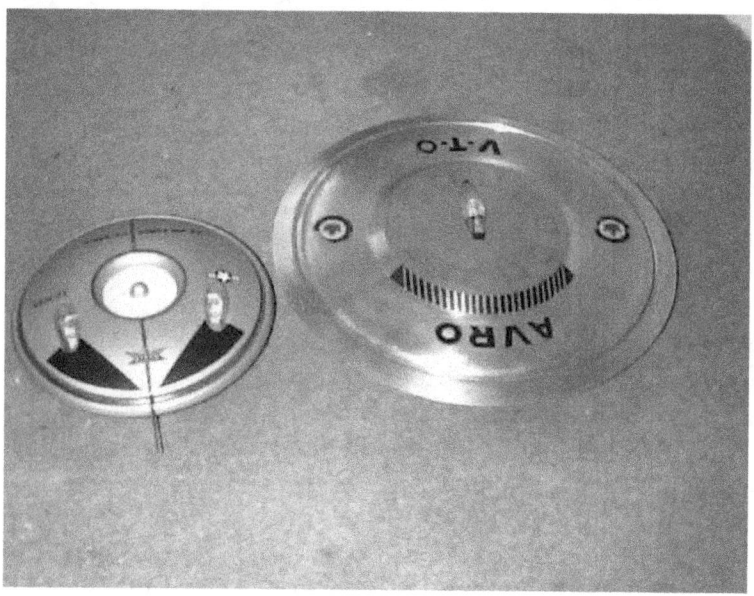

Fig. 9. Avro company models of the Y-2 (right) and the Avrocar (left).

A wide variety of designs were studied for a VTOL fighter aircraft, all revolved around the disk shape, leading to the Project 1794 involving a supersonic large disk fighter aircraft. The concept proceeded to wind tunnel testing with a variety of scale models. It featured a raised section in the middle over the engine, the intake covered with a series of louvers that would be closed in forward flight. Frost's performance estimates for the concept were for a potential of Mach 3.5 at 100,000 ft (30,000 m) altitudes.

There was some debate about the concept within the USAF, as many groups were attempting to gain funding for their own pet projects, like nuclear powered bombers. In a repeat of the earlier Toronto Star release, in 1955 an extensive article appeared in Look Magazine that, among other claims, speculated that current UFO sightings were Soviet-built saucers. The article went on to describe such an aircraft with diagrams that were clearly influenced by the Avro design.

A new impeller-driven engine design was proposed as Avro PV-704 (PV stood for Private Venture), powered by six Armstrong Siddeley Viper jet engines blowing across the outer rim of a central rotor. The PV-704 was a "stop-gap" design built into a bunker-like building behind the Avro Experimental Test facility. It was intended to test various Project 1794 concepts and provide the USAF with test data to show the viability of the concept.

The original plan to initially test the "Viper Engine Rig" was to have continued into "free flight" testing. Unfortunately, testing was anything but smooth; the test model suffered from hazardous oil leaks, resulting in three fires.

It eventually got to the point that staff were afraid of the machine, even when safely ensconced in a booth constructed of bullet-proof glass and quarter-inch-thick steel. A final, disastrous and nearly lethal engine test in 1956 which involved a Viper jet engine "running wild" convinced Frost that a less dangerous test vehicle was necessary.

To gather flight data on the basic concept while the engine development continued, in 1958 Frost proposed building a smaller "proof-of-concept" test vehicle he called the Avrocar. By this point, the US Army was involved in a wide variety of experiments on smaller VTOL aircraft that would act as a "flying Jeep," and they became interested in Avro's concept as well. Frost pitched his smaller design both as a prototype of a vehicle suitable for the Army's needs, as well as an aerodynamic testbed for the WS-606. Initial performance requirements for the Avrocar were a ten-minute hover capability in ground effect and 25-mile (40 km) range with a 1,000 lb (450 kg) payload.

The new plan appeared to make everybody happy, and a $2 million joint-services contract managed by the Air Force was awarded to Avro to build and test two Avrocars, which the Army referred to as the VZ-9-AV (with AV standing for "Avro," an unusual departure from normal US Army nomenclature), the latest in a series of "VZ" aircraft. Army interest in the Avrocar program was apparently very high. Bernard Lindenbaum recalls a trip to Washington in the late 1950s to request additional funding for a study on helicopter drag reduction.

Although the funding was approved, he overheard an Army General remark that the Huey would be the last helicopter the Army would buy since the helicopter would be replaced by the Avrocar.

Additional Air Force funding of approximately $700,000 (unexpended from the 606A program) was also moved to the Avrocar project. In March 1959, an additional $1.77 million contract was received for a second prototype. At rollout, projected performance was far in excess of the requirement, with a 225 knots (417 km/h) maximum speed, 10,000 feet (3,000 m) ceiling, 130-mile (209 km) range with 1,000 lb (450 kg) payload, and hover out of ground effect with 2,428 lb (1,101 kg) payload.

Maximum takeoff weight with transition to forward flight out of ground effect was calculated to be 5,650 lb (2,560 kg), maximum weight with a transition in ground effect (GETOL) was 6,970 lb (3,160 kg).

Just as the first working test models were being manufactured, disaster struck. The Canadian government cancelled the Avro CF-105 Arrow program on "Black Friday," 20 February 1959. The ensuing result was the lay-off of almost all Avro Canada employees, including those with the Special Projects Group.

However, three days following the announcement of the Arrow cancellation, many of the Special Projects employees were rehired. But it wasn't quite business as usual. The team now included people from the CF-100 and CF-105 teams and the Special Projects Group was moved into the main building, which was nearly empty. As well, company "brass" became more involved in the group's operations.

The USAF Project Office devoted to the Avro projects, recommended that the WS-606A and all related work (including the Avrocar) be cancelled. A "stop/go" work order came down and Frost was forced once more to try and rescue the project. In an elaborate effort, Frost made a resounding case for continuation of US military funding. Late in May 1959, the USAF authorized Avro to continue the "flying saucer" programs (fig. 10).

Fig. 10. US Army Avrocars depicted as "flying jeeps" in company literature.

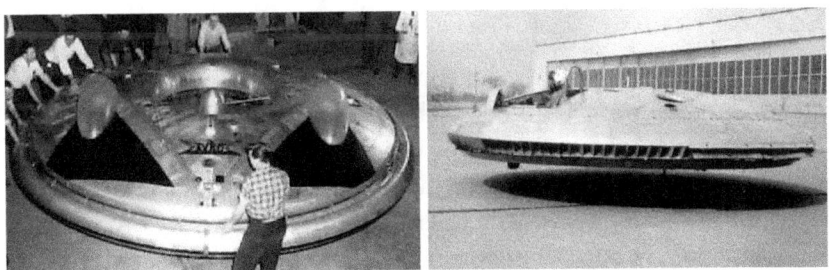

Fig. 11. First Avrocars 1958 (left); the second model 1960 (right).

The Avro VZ-9 Avrocar was a "dead end" in VTOL design (fig. 11), according to Russell Lee, curator at the National Air and Space Museum, yet its technological innovations have intrigued other designers. One of the design elements it embodied, the use of ducted fans led to other experimental programs. Dr. Paul Moller, a Canadian expatriate who had worked at Avro Canada as a young engineer, based an initial series of experimental VTOL vehicles on "saucer" technology utilizing the buried ducted fan à la-Avrocar. The XM-2, the first of the series looked remarkably like a miniature flying saucer. After successful tether tests, the saucer designs also at one time publicized as "discojet" were abandoned and their latest project, the Moller Skycar, has a flying-car appearance.

NASA's Aeronautics Research Mission Directorate (ARMD) works to solve the challenges that still exist in our nation's air transportation system: air traffic congestion, safety and environmental impacts.

Solutions to these problems require innovative technical concepts, and dedicated research and development. NASA's ARMD pursues the development of new flight operation concepts, and new tools and technologies that can transition smoothly to industry to become products.

Through green aviation, NASA is helping create safer, greener and more effective travel for everyone. Our green aviation goals are to enable fuel-efficient flight planning, and reduce aircraft fuel consumption, emissions and noise.

NASA aeronautics' four research programs conduct fundamental, cutting-edge research into new aircraft technologies, as well as systems-level research into the integration of new operations concepts and technologies into the Next Generation Air Transportation System (NextGen). A fifth program manages a portfolio of wind tunnels and other testing facilities (icing, propulsion), flight research and support aircraft, and the evolution of test technologies at NASA centers around the country.

The Dryden Flight Research Center is NASA's primary center for atmospheric flight research and operations. NASA Dryden is critical in carrying out the agency's missions of space exploration, space operations, scientific discovery, and aeronautical research and development (R&D).

Located at Edwards, California, in the western Mojave Desert, Dryden is uniquely situated to take advantage of the excellent year-round flying weather, remote area, and visibility to test some of the nation's most exciting air vehicles.

In support of space exploration, we are managing the launch abort systems testing and integration, in partnership with the Johnson Space Center and Lockheed Martin, for the Crew Exploration Vehicle that will replace the Space Shuttle.

Dryden is the primary alternate landing site for the Space Shuttle and orbital support for the International Space Station.

In support of scientific discovery, we are managing the Stratospheric Observatory for Infrared Astronomy (SOFIA) program - a flying telescope aboard a Boeing 747 aircraft - in partnership with the Ames Research Center and the German Aerospace Center.

In support of aeronautical R&D, we are involved in many aspects of the Fundamental Aeronautics and Aviation Safety programs, including the X-48 Blended Wing Body and Ikhana (Predator B) in support of subsonics and Adaptive Flight Controls in support of the Aviation Safety Program.

For 60 years, Projects at Dryden have led to major advancements in the design and capabilities of many state-of-the-art civilian and military aircraft. The newest, the fastest, the highest - all have made their debut in the vast, clear desert skies over Dryden.

Dryden Flight Research Center plays a vital role in advancing technology and science through flight. Here, we demonstrate America's leadership in aeronautics and space technology as we continue to push the envelope to revolutionize aviation and pioneer aerospace technology.

NASA is operating two Lockheed ER-2 Earth resources aircraft as flying laboratories in the Sub-Orbital Science Program under the agency's Science Mission Directorate. The aircraft, based at NASA's Dryden Flight Research Center, Edwards, Calif., collect information about our surroundings, including Earth resources, celestial observations, atmospheric chemistry and dynamics, and oceanic processes. The aircraft also are used for electronic sensor research and development, satellite calibration and satellite data validation (fig. 12).

Fig. 12. *Lockheed ER-2*

The F-15B Research Testbed is a modified twin-engine jet fighter that provides NASA, industry, and universities with long-term capability for the efficient flight test of aerodynamic, instrumentation, propulsion, and other flight research experiments. This aircraft is a unique airborne resource, and is considered by researchers to be a virtual "flying wind tunnel" and reliable supersonic testbed. In addition to flying research missions, Dryden's F-15B also is used for crew training, pilot proficiency, and safety chase support for other research aircraft.

Bearing NASA tail number 836, the F-15B is about 64 feet long and has a wingspan of just under 43 feet. It is powered by two Pratt and Whitney F100-PW-100 turbofan engines that can produce almost 24,000 pounds of thrust each in full afterburner. It is capable of dash speeds of Mach 2.3, or 2.3 times the speed of sound, at altitudes of 40,000 to 60,000 feet. With an external flight test fixture mounted beneath the fuselage in place of the standard external fuel tank, speeds are limited to Mach 2.0. The aircraft has a full-fuel takeoff weight of about 42,000 pounds and a landing weight of about 32,000 pounds. It has aerial refueling capability for extended-duration research missions.

The data acquisition system in the aircraft makes the F-15B one of the most versatile testbed aircraft NASA flies. An on-board video system monitored from the rear seat of the cockpit provides a high-speed airborne video and photo capability that can be downlinked to researchers on the ground. The data system includes a research airdata system for the aircraft, as well as a Global Positioning System (GPS) navigation package; a radome with a nose boom that contains an airdata probe; a digital data recorder; and telemetry antennas.

Recent Activity In 2008, 836 hosted several research projects aimed understanding and overcoming the challenges associated with providing civilian overland supersonic transport as well as advanced propulsion system design.

Lift and Nozzle Change Effects on Tail Shock (LaNCETS) quantified how changes in lift distribution and nozzle area ratio affect the supersonic shock structure on the aft end of NASA's highly modified F-15 (837) research aircraft.

The Propulsion Flight Test Fixture (PFTF) was flown to quantify the flow field surrounding a research inlet (to be flown in early 2009). This project brings the PFTF one step closer to full operational capability, which will allow the F-15B to demonstrate and study advanced propulsion concepts in flight.

In 2006, Gulfstream Aerospace and Dryden teamed in a project called Quiet SpikeTM to investigate the suppression of sonic booms. The project centered around a retractable, 24-foot-long lance-like spike mounted on the nose of NASA Dryden's F-15B (#836) research testbed aircraft. The spike, made primarily of composite materials, created three small shock waves that traveled parallel to each other all the way to the ground, producing less noise than typical shock waves that build up at the front of supersonic jets.

This highly successful project put spike-induced sonic boom suppression theory to the test in the actual flight environment afforded by NASA's supersonic F-15B (fig. 13).

Fig. 13. The supersonic F-15B.

NASA's Dryden Flight Research Center has acquired two developmental Northrop Grumman autonomously operated Global Hawk aircraft for use in high-altitude, long-duration Earth science missions. Global Hawk measures 44 feet in length, with a wingspan of 116 feet. NASA expects to operate the Global Hawk with payloads up to 2000 pounds and at altitudes up to 65, 000 feet. Its range is greater than 10,000 nautical miles and its endurance is greater than 31 hours (fig. 14).

Fig. 14. The Global Hawk.

NASA's Ikhana unmanned aircraft completed the fourth in a series of wildfire imaging demonstration flights Thursday over central and southern California. The 10-hour flight departed NASA's Dryden Flight Research Center at Edwards Air Force Base shortly after 6 a.m. PDT Thursday, and returned about 4 p.m. (fig. 15).

The flight was part of the Western States Fire Mission being conducted by NASA and the U.S. Forest Service, which is demonstrating improved wildfire imaging and mapping capabilities of a sophisticated thermal-infrared imaging sensor and real-time data communications equipment developed at NASA's Ames Research Center. Prior flights lasting as long as 20 hours in August and

September took the Ikhana and its imaging payload over a number of wildfires in southern and central California, Oregon, Idaho, Washington, western Wyoming and southwestern Montana.

Thursday's flight included imaging previously burned areas in central and southern California for a Burned Area Rehabilitation and Stabilization (BAER) assessment while the Ikhana cruised at about 23,000 feet altitude in the national airspace. The aircraft then re-entered the restricted military test range airspace near Edwards and climbed to over 40,000 feet altitude in an evaluation of the Ikhana's high-altitude performance capabilities with the 420-lb. instrument pod tucked under its left wing.

Fig. 15. Ikhana unmanned aircraft.

NASA Dryden Pad Abort-1 flight test team technicians Kevin Mount and Jeff Doughty prepare to roll the platform carrying the Orion flight test crew module from the cavernous interior of a C-17 cargo plane following its return to NASA Dryden Flight Research Center on June 15, 2010 (fig. 16). The boilerplate module was undamaged in the PA-1 test of the Launch Abort System at White Sands Missile Range on May 6, 2010.

The undamaged Pad Abort-1 flight test crew module rests in the desert after a successful Pad Abort-1 flight test May 6 at the White Sands Missile Range in New Mexico.

Officials and media representatives watch the launch of an Orion test crew module and its launch abort system during NASA's Pad Abort 1 test on May 6 at White Sands Missile Range, N.M. PA-1 was the first fully integrated flight test of the crew rescue system being developed for the Orion crew capsule.

Fig. 16. NASA Dryden Pad Abort-1 flight test team technicians Kevin Mount and Jeff Doughty prepare to roll the platform carrying the Orion flight test crew module from the cavernous interior of a C-17 cargo plane.

Boeing Phantom Works has partnered with NASA and the Air Force Research Laboratory to study the structural, aerodynamic and operational advantages of the Blended Wing Body concept, a cross between a conventional plane and a flying wing design. The Air Force has designated the prototype the X-48B based on its interest in the design's potential as a multi-role, long-range, high-capacity military transport aircraft. The 8.5 percent scale, remotely piloted X-48B is dynamically scaled to fly much like the full-size aircraft would fly (fig. 17).

Following completion of installation of test instrumentation, one of two X-48B Blended Wing Body technology demonstrators began flight tests at NASA's Dryden Flight Research Center in early 2007, and those tests are continuing into 2008. Researchers at NASA's Langley Research Center in Hampton, Va., tested the second X-48B prototype aircraft in Langley's historic full-scale wind tunnel in the spring of 2006, and the flight tests are intended in part to validate the results of those wind tunnel tests.

Advantages of the blended wing-body concept include high fuel efficiency, low noise and a large payload volume for the size of the aircraft. Flight testing at NASA Dryden will focus on the low-speed, low-altitude flight characteristics of the blended wing-body configuration, including engine-out control, stall characteristics and handling qualities. The short flight test program is intended to demonstrate that the novel design can be flown as safely as current transports having a traditional fuselage, wings and tail configuration.

Fig. 17. X-48B Blended Wing Body.

The two X-48B Blended Wing Body technology demonstration aircraft were built by Cranfield Aerospace in the United Kingdom to Boeing's specifications. The subscale prototypes have a wingspan of 20.4 feet, with prominent vertical fins and rudders at the wingtips and elevons along the trailing edges of the wings. The 523-lb. gross weight aircraft are powered by three small model aircraft turbojet engines providing a maximum combined thrust of about 160 lbs. The X-48B has an estimated top airspeed of 118 knots (139 mph), a maximum altitude of about 10,000 feet and a flight duration of about 40 minutes.

According to NASA, "ERAST is a multiyear effort to develop the aeronautical and sensor technologies for a new family of remotely piloted aircraft intended for upper atmospheric science missions. Designed to cruise at slow speeds for long durations at altitudes of 60,000 to 100,000 ft, such aircraft could be used to collect, identify, and monitor environmental data to assess global climate change and assist in weather monitoring and forecasting. They also could serve as airborne telecommunications platforms, performing functions similar to communications satellites at a fraction of the cost of lofting a satellite into space." (fig. 18).

"The ERAST program is sponsored by the Office of Aeronautics and Space Transportation Technology at NASA Headquarters, and is managed by NASA Dryden Flight Research Center. The NASA Ames Research Center, Moffett Field, California, heads the sensor technology development. The NASA Lewis Research Center, Cleveland, Ohio, and NASA Langley Research Center, Hampton, Virginia, are contributing expertise in the areas of propulsion, structures, and systems analysis. Several small high-technology aeronautical development firms, including ALTUS developer General Atomics Aeronautical Systems, Inc., are teamed with NASA in the ERAST Alliance to work towards common goals of the program." Industry partners in the ERAST Alliance included Aurora Flight Systems, AeroVironment, General Atomics, Scaled Composites, Thermo-Mechanical Systems, Hyperspectral Sciences, and Longitude 122 West.

The types of science mission which ERAST prepares for can include remote sensing for Earth sciences studies, hyperspectral imaging for agriculture monitoring, tracking of severe storms, and serving as telecommunications relay platforms.

A parallel effort headed by Ames developed lightweight, microminiaturized sensors that can be carried by these aircraft for environmental research and Earth monitoring.

Additional technologies considered by the ERAST Alliance include lightweight materials, avionics, aerodynamics, and other forms of propulsion suitable for extreme altitudes and duration.

Although ERAST Alliance members were responsible for aircraft development and operation, NASA had primary responsibility for overall program leadership, major funding, individual project management, development and coordination of payloads. NASA also worked on long-term issues with the Federal Aviation Administration and developed technology to make operation of these remotely operated aircraft in national airspace practical.

Fig. 18. ERAST is a multiyear effort to develop the aeronautical and sensor technologies for a new family of remotely piloted aircraft intended for upper atmospheric science missions.

The Lockheed SR-71 "Blackbird" (fig. 19) was an advanced, long-range, Mach 3+ strategic reconnaissance aircraft. It was developed as a black project from the Lockheed A-12 reconnaissance aircraft in the 1960s by the Lockheed Skunk Works. Clarence "Kelly" Johnson was responsible for many of the design's innovative concepts. During reconnaissance missions the SR-71 operated at high speeds and altitudes to allow it to outrace threats. If a surface-to-air missile launch was detected, the standard evasive action was simply to accelerate and outrun the missile.

The SR-71 served with the U.S. Air Force from 1964 to 1998. Although twelve of the 32 aircraft built were destroyed in accidents, none were lost to enemy action. The SR-71 was unofficially named the Blackbird, and called the Habu by its crews, referring to an Okinawan species of pit viper. Since 1976, it has held the world record for the fastest air-breathing manned aircraft, a record previously held by the YF-12.

Fig. 19. *Lockheed SR-71 Blackbird.*

Shuttle Carrier Aircraft

NASA uses two modified Boeing 747 jetliners, originally manufactured for commercial use, as Space Shuttle Carrier Aircraft

(SCA). One is a 747-100 model, while the other is designated a 747-100SR (short range). The two aircraft are identical in appearance and in their performance as Shuttle Carrier Aircraft (Fig. 20).

The 747 series of aircraft are four-engine intercontinental-range, swept-wing "jumbo jets" that entered commercial service in 1969.

The SCAs are used to ferry space shuttle orbiters from landing sites back to the launch complex at the Kennedy Space Center and also to and from other locations too distant for the orbiters to be delivered by ground transportation. The orbiters are placed atop the SCAs by Mate-Demate Devices, large gantry-like structures that hoist the orbiters off the ground for post-flight servicing and then mate them with the SCAs for ferry flights.

Fig. 20. NASA uses two modified Boeing 747 jetliners, originally manufactured for commercial use, as Shuttle Carrier Aircraft (SCA).

NASA 905

NASA 905 was the first SCA. It was obtained from American Airlines in 1974. Shortly after acceptance by NASA, the SCA flew a series of wake vortex research flights at the Dryden Flight Research Center, Edwards, Calif., in a study to seek ways of reducing turbulence produced by large aircraft. Pilots flying as much as several miles behind large aircraft have encountered wake turbulence that has caused control problems. The NASA study helped the Federal Aviation Administration modify flight procedures for commercial aircraft during airport approaches and departures.

Following the wake vortex studies, NASA 905 was modified by Boeing to its present SCA configuration and the aircraft was returned to Dryden for its role in the 1977 Space Shuttle Approach and Landing Tests (ALT). This series of eight captive and five free flights with the orbiter prototype Enterprise, in addition to ground taxi tests, validated the aircraft's performance as an SCA, in addition to verifying the glide and landing characteristics of the orbiter configuration — paving the way for orbital flights.

A flight crew escape system, consisting of an exit tunnel extending from the flight deck to a hatch in the bottom of the fuselage, was installed during the modifications. The system also included a pyrotechnic system to activate the hatch release and cabin window release mechanisms. The flight crew escape system was removed from the NASA 905 following the successful completion of the ALT program.

NASA 905 was the only SCA used by the space shuttle program until November 1990, when NASA 911 was delivered as an SCA. Along with ferrying Enterprise and the flight rated orbiters between the launch and landing sites and other locations, NASA 905 also ferried Enterprise to Europe for display in England and at the Paris Air Show.

NASA 911

The second SCA is designated NASA 911. It was obtained by NASA from Japan Airlines (JAL) in 1989. It was also modified by Boeing Corporation. It was delivered to NASA on Nov. 20, 1990.

The Space Shuttle prototype Enterprise flies free after being released from NASA's 747 Shuttle Carrier Aircraft (SCA) over Rogers Dry Lake during the second of five free flights carried out at the Dryden Flight Research Center, Edwards, California, as part of the Shuttle program's Approach and Landing Tests (ALT) in 1977 (fig. 21).

The tests were conducted to verify orbiter aerodynamics and handling characteristics in preparation for orbital flights with the Space Shuttle Columbia. A tail cone over the main engine area of Enterprise smoothed out turbulent airflow during flight. It was removed on the two last free flights to accurately check approach and landing characteristics. A series of test flights during which Enterprise was taken aloft atop the SCA, but was not released, preceded the free flight tests.

Fig. 21. *The Space Shuttle prototype Enterprise flies free after being released from NASA's 747 Shuttle Carrier Aircraft.*

The Space Shuttle prototype Enterprise flies free of NASA's 747 Shuttle Carrier Aircraft (SCA) during one of five free flights carried out at the Dryden Flight Research Facility, Edwards, California in 1977 as part of the Shuttle program's Approach and Landing Tests (ALT). The tests were conducted to verify orbiter aerodynamics and handling characteristics in preparation for orbital flights with the Space Shuttle Columbia. A tail cone over the main engine area of Enterprise smoothed out turbulent airflow during flight. It was removed on the two last free flights to accurately check approach and landing characteristics (fig. 22).

Fig. 22. The Space Shuttle prototype Enterprise flies free.

The Boeing B-52 Stratofortress is a long-range, subsonic, jet-powered strategic bomber designed and built by Boeing and operated by the United States Air Force (USAF).

Beginning with the successful contract bid on 5 June 1946, the B-52 design evolved from a straight-wing aircraft powered by six turboprop engines to the final prototype YB-52 with eight turbojet engines and swept wings. The Stratofortress took its maiden flight in April 1952. Built to carry nuclear weapons for Cold War-era deterrence missions, the B-52 Stratofortress replaced the Convair B-36. Although a veteran of a number of wars, the Stratofortress has dropped only conventional munitions in combat. The B-52 carries up to 70,000 pounds (32,000 kg) of weapons. Its Stratofortress name is rarely used outside of official contexts; it has been referred to by Air Force personnel as the BUFF (Big Ugly Fat/Flying Fucker/Fellow).

The B-52 has been in active service with the USAF since 1955. The bombers flew under the Strategic Air Command (SAC) until it was disestablished in 1992 and its aircraft absorbed into the

Air Combat Command (ACC). This remained the case until February 2010 when all B-52 Stratofortress and B-2 Spirit aircraft were transferred from ACC to the recently established Air Force Global Strike Command (AFGSC). Superior performance at high subsonic speeds and relatively low operating costs have kept the B-52 in service despite the advent of later aircraft, including the Mach-3 North American XB-70 Valkyrie, the supersonic Rockwell B-1B Lancer, and the stealthy Northrop Grumman B-2 Spirit. The B-52 marked its 50th anniversary of continuous service with its original primary operator in 2005. (Other aircraft with similarly long service include the English Electric Canberra, Tupolev Tu-95, Lockheed C-130 Hercules, Boeing KC-135 Stratotanker, and Lockheed U-2, fig. 23)

Fig. 23. A B-52H from Barksdale AFB flying over the desert.

In May 1948, AMC asked Boeing to incorporate the previously discarded, but now more fuel-efficient, jet engine into the design. This resulted in Boeing developing yet another revision – in July 1948, Model 464-40 substituted Westinghouse J40 turbojets for the turboprops. Nevertheless, on 21 October 1948, Boeing was told to create an entirely new aircraft using Pratt & Whitney J57 turbojets.

On 25 October, Boeing engineers produced a proposal and a hand-carved model of 464-49. The new design built upon the basic layout of the B-47 Stratojet with 35° swept wings, eight engines paired in four underwing pods, and bicycle landing gear with wingtip outrigger wheels. A notable feature of the landing gear was the ability to pivot the main landing gear up to 20° from the aircraft centerline to increase safety during crosswind landings. The aircraft was projected to exceed all design specifications. Although the full-size mock-up inspection in April 1949 was generally favorable, range again became a concern since the J40s and early model J57s had excessive fuel consumption.

Despite talk of another revision of specifications or even a full design competition among aircraft manufacturers, General LeMay, now in charge of Strategic Air Command, insisted that performance should not be compromised due to delays in engine development. In a final attempt to increase range, Boeing created the larger 464-67, stating that once in production, the range could be further increased in subsequent modifications.

Following several direct interventions by LeMay, Boeing was awarded a production contract for 13 B-52As and 17 detachable reconnaissance pods on 14 February 1951. The last major design change, also at the insistence of General LeMay, was a switch from the B-47 style tandem seating to a more conventional side-by-side cockpit which increased the effectiveness of the copilot and reduced crew fatigue. Both XB-52 (fig. 24) prototypes featured the original tandem seating arrangement with a framed bubble-type canopy.

The YB-52, the second XB-52 modified with more operational equipment, first flew on 15 April 1952 with "Tex" Johnston as pilot.[Notes 1] A 2 hour, 21-minute proving flight from Boeing Field, King County, near Renton, Washington to Larson AFB was undertaken with Boeing test pilot Alvin M. Johnston and Air Force Lieutenant Colonel Guy M.

Townsend. The XB-52 followed on 2 October 1952. The thorough development, including 670 days in the wind tunnel and 130 days of aerodynamic and aeroelastic testing, paid off with smooth flight testing. Encouraged, the Air Force increased its order to 282 B-52s.

Fig. 24. *A XB-52 Prototype on flight line (X-4 in foreground).*

The YF-17 (fig. 25) was primarily constructed of aluminum, in conventional semi-monocoque stressed-skin construction, though over 900 lb (408 kg) of its structure were graphite/epoxy composite. The small nose contained a simple ranging radar. The cockpit sported an ejection seat inclined at 18°, a bubble canopy, and a head-up display (HUD).

The thin wings carried no fuel, and in areas such as the leading and trailing edge and the LERX, were composed of a Nomex honeycomb core with composite facesheets.

The rear of the aircraft featured twin all-moving stabilators of aluminum over a honeycomb core, and twin vertical stabilizers of a conventional construction.

Like the wings, the leading and trailing edges were constructed of composite facesheets over honeycomb core. A composite speedbrake was located above and between the engines.

The aircraft was powered by a pair of 14,400-pound-force (64 kN) General Electric YJ101-GE-100 turbofans, a development of the GE15, mounted next to each other to minimize thrust asymmetry in the event of an engine loss. For ease of maintenance, the engines are mounted in a steady-rest that allows removal from below the aircraft, without disturbing the empennage controls.

Each engine drove an independent hydraulic system. Unlike the P-530, the YF-17 sported a partial fly-by-wire control scheme, formally called the electronic control augmentation system (ECAS), utilizing ailerons, rudders, and stabilators for primary flight control.

Fig. 25. YF-17 Cobra, operated by NASA.

The McDonnell Douglas F-4 Phantom II (fig. 26) started as the prototype McDonnell XF4H-1, which made its first flight on May 27, 1958 with crack test pilot Robert Little at the controls.

After five years of strenuous effort, a happy combination of brains, determination, timing, and luck enabled the McDonnell Corporation to deliver an aircraft that leapfrogged a generation of fighter aircraft in development.

Fig. 26. F-4 Phantom II.

In 1966 the U.S. Air Force formed the Attack Experimental (A-X) program office. On 6 March 1967, the Air Force released a request for information to 21 defense contractors for the A-X.

The objective was to create a design study for a low-cost attack aircraft. The officer in charge of the project was Colonel Avery Kay.

In 1969, the Secretary of the Air Force asked Pierre Sprey to write the detailed specifications for the proposed A-X project. However, his initial involvement was kept secret because of Sprey's earlier controversial involvement in the F-X project.

Sprey's discussions with A-1 Skyraider pilots operating in Vietnam and analysis of the effectiveness of current aircraft used in the role indicated the ideal aircraft should have long loiter time, low-speed maneuverability, massive cannon firepower, and extreme survivability; an aircraft that had the best elements of the Ilyushin Il-2, Henschel Hs 129 and Skyraider.

The specifications also demanded that each aircraft cost less than $3 million. Sprey required that the biography of World War II attack pilot Hans-Ulrich Rudel be read by people on A-X program.

In May 1970, the USAF issued a modified and much more detailed request for proposals (RFP) for the aircraft.

The threat of Soviet armored forces and all-weather attack operations had became more serious.

Now included in the requirements was that the aircraft would be designed specifically for the 30 mm cannon.

The RFP also specified an aircraft with a maximum speed of 460 mph (740 km/h), takeoff distance of 4,000 feet (1,200 m), external load of 16,000 pounds (7,300 kg), 285-mile (460 km) mission radius, and a unit cost of US$1.4 million.

During this time, a separate RFP was released for A-X's 30 mm cannon with requirements for a high rate of fire (4,000 round/minute) and a high muzzle velocity.

Six companies submitted proposals to the USAF, with Northrop and Fairchild Republic selected to build prototypes: the YA-9A and YA-10A, respectively.

General Electric and Philco-Ford were selected to build and test GAU-8 cannon prototypes.

The YA-10A first flew on 10 May 1972. After trials and a fly-off against the YA-9A, the Air Force announced its selection of Fairchild-Republic's YA-10A on 18 January 1973 for production.

General Electric was selected to build the GAU-8 cannon in June 1973. The YA-10 had an additional fly-off in 1974 against the Ling-Temco-Vought A-7D Corsair II, the principal Air Force attack aircraft at the time, in order to prove the need to purchase a new attack aircraft. The first production A-10 flew in October 1975, and deliveries to the Air Force commenced in March 1976. In total, 715 airplanes were produced, the last delivered in 1984.

Fig. 27. A-10 Thunderbolt.

One experimental two-seat A-10 Night Adverse Weather (N/AW) version was built by converting an A-10A. The N/AW was developed by Fairchild from the first Demonstration Testing and

Evaluation (DT&E) A-10 (fig. 27) for consideration by the USAF. It included a second seat for a weapons system officer responsible for electronic countermeasures (ECM), navigation and target acquisition. The variant was canceled, and the only two-seat A-10 built now resides at Edwards Air Force Base's Flight Test Center Museum. The N/AW version did not interest the USAF or export customers. The two-seat trainer version was ordered by the Air Force in 1981, but funding was canceled by U.S. Congress and the jet was not produced.

The Embraer EMB 314 Super Tucano, also named ALX or A-29 (fig. 28) is a turboprop aircraft designed for light attack, counter insurgency (COIN) and pilot training missions, incorporating modern avionics and weapons systems. It is currently in service with the air forces of Brazil, Dominican Republic and Colombia, and has been ordered by the forces of Chile and Ecuador. Embraer has plans to sell it to other countries in Asia and the Middle East. Besides pilot training, it is heavily employed in monitoring operations within the Amazon basin.

Fig. 28. *A-29 Super Tucano.*

The Skyhawk was designed by Douglas Aircraft's Ed Heinemann in response to a U.S. Navy call for a jet-powered attack aircraft to replace the older AD Skyraider.

Heinemann opted for a design that would minimize its size, weight, and complexity.

The result was an aircraft that weighed only half of the Navy's weight specification. It had a wing so compact that it did not need to be folded for carrier stowage.

The diminutive Skyhawk soon received the nicknames "Scooter", "Kiddiecar", "Bantam Bomber", "Tinker Toy Bomber", and, on account of its nimble performance, "Heinemann's Hot-Rod".

The aircraft is of conventional post-World War II design, with a low-mounted delta wing, tricycle undercarriage, and a single turbojet engine in the rear fuselage, with two air intakes on the fuselage sides. The tail is of cruciform design, with the horizontal stabilizer mounted above the fuselage. Armament consisted of two 20 mm (.79 in caliber) Colt Mk 12 cannons, one in each wing root, with 200 rpg, plus a large variety of bombs, rockets, and missiles carried on a hardpoint under the fuselage centerline and hardpoints under each wing (originally one per wing, later two).

The choice of a delta wing, for example, combined speed and maneuverability with a large fuel capacity and small overall size, thus not requiring folding wings, albeit at the expense of cruising efficiency.

The leading edge slats were designed to drop automatically at the appropriate speed by gravity and air pressure, saving weight and space by omitting actuation motors and switches.

Similarly the main undercarriage did not penetrate the main wing spar, designed so that when retracted only the wheel itself was inside the wing and the undercarriage struts were housed in a fairing below the wing.

The wing structure itself could be lighter with the same overall strength and the absence of a wing folding mechanism further reduced weight.

This is the opposite of what can often happen in aircraft design where a small weight increase in one area leads to a compounding increase in weight in other areas to compensate, leading to the need for more powerful, heavier engines and so on in a vicious circle.

The A-4 pioneered the concept of "buddy" air-to-air refueling. This allows the aircraft to supply others of the same type, eliminating the need of dedicated tanker aircraft—a particular advantage for small air arms or when operating in remote locations.

This allows for greatly improved operational flexibility and reassurance against the loss or malfunction of tanker aircraft, though this procedure reduces the effective combat force on board the carrier.

A designated supply A-4 would mount a center-mounted "buddy store", a large external fuel tank with a hose reel in the aft section and an extensible drogue refueling bucket.

This aircraft was fueled up without armament and launched first.

Attack aircraft would be armed to the maximum and given as much fuel as was allowable by maximum takeoff weight limits, far less than a full tank.

Once airborne, they would then proceed to top off their fuel tanks from the tanker using the A-4's fixed refueling probe on the starboard side of the aircraft nose. They could then sortie with both full armament and fuel loads.

While rarely used in U.S. service since the KA-3 Skywarrior tanker became available, the F/A-18E/F Super Hornet includes this capability.

The A-4 was also designed to be able to make an emergency landing, in the event of a hydraulic failure, on the two drop tanks nearly always carried by these aircraft. Such landings resulted in only minor damage to the nose of the aircraft which could be repaired in less than an hour.

The Navy issued a contract for the type on 12 June 1952, and the first prototype first flew from Edwards Air Force Base, California on 22 June 1954. Deliveries to Navy and Marine Corps squadrons (to VA-72 and VMA-224 respectively) commenced in late 1956.

The Skyhawk remained in production until 1979, with 2,960 aircraft built, including 555 two-seat trainers.

The last production A-4, an A-4M issued to a Marine squadron (VMA-223) had the flags of all nations who had operated the A-4 series aircraft painted on the fuselage sides.

The Skyhawk proved to be a relatively common United States Navy aircraft export of the postwar era. Due to its small size, it could be operated from the older, smaller World War II-era aircraft carriers still used by many smaller navies during the 1960s.

These older ships were often unable to accommodate newer Navy fighters such as the F-4 Phantom II and F-8 Crusader, which were faster and more capable than the A-4, but significantly larger and heavier than older naval fighters.

The Navy operated the A-4 in both Regular Navy and Naval Reserve light attack squadrons (VA). Although the A-4's use as a training and adversary aircraft would continue well into the 1990s, the Navy began removing the aircraft from its front line attack squadrons in 1967, with the last ones (Super Foxes of VA-55/212/164) being retired in 1976.

The Marine Corps would not take the U.S. Navy's replacement warplane, the A-7 Corsair II, instead keeping Skyhawks in service with both Regular Marine Corps and Marine Corps Reserve attack squadrons (VMA), and ordering the new A-4M model. The last USMC Skyhawk was delivered in 1979, and they were used until the mid-1980s before they were replaced by the equally small, but more versatile STOVL AV-8 Harrier II.

VMA-131, Marine Aircraft Group 49 (the Diamondbacks) retired its last four OA-4Ms on 22 June 1994. Lieutenant Colonel George "Eagle" Lake III (CO), Major John "Baja" Rufo (XO), Captain Dave "Yoda" Hurston, and Major Mike "Struts" Volland flew a final official USMC A-4 sortie during the A-4 Standdown Ceremony.

Trainer versions of the Skyhawk remained in Navy service, however, finding a new lease on life with the advent of "adversary training", where the nimble A-4 was used as a stand-in for the Mikoyan-Gurevich MiG-17 in dissimilar air combat training (DACT). It served in that role at "Top Gun" until 1999.

The A-4's nimble performance also made it suitable to replace the F-4 Phantom II when the Navy downsized its aircraft for the Blue Angels demonstration team, until F/A-18 Hornets were available in the 1980s.

The last U.S. Navy Skyhawks, TA-4J models belonging to the composite squadron VC-8, remained in military use for target-towing, and as adversary aircraft, for combat training at Naval Station Roosevelt Roads. These aircraft were officially retired on 3 May 2003.

Skyhawks were well-loved by their crews for being tough and agile.

These attributes, along with their low purchase and operating cost as well as easy maintenance, have contributed to the popularity of the A-4 with American and international armed forces. Besides the United States, at least three other nations have used A-4 Skyhawks in combat (Argentina, Israel, and Kuwait).

Skyhawks were the Navy's primary light bomber used over North Vietnam during the early years of the Vietnam War while the USAF was flying the supersonic F-105 Thunderchief; they were later supplanted by the A-7 Corsair II in the Navy light bomber role. Skyhawks carried out some of the first air strikes by the US during the conflict, and a Marine Skyhawk is believed to have dropped the last American bombs on the country.

Notable naval aviators who flew the Skyhawk included Lieutenant Commanders Everett Alvarez Jr. and John McCain, and Commander James Stockdale. On 1 May 1967, an A-4C Skyhawk piloted by Lieutenant Commander Theodore R. Swartz of VA-76 aboard the carrier USS Bon Homme Richard, shot down a North Vietnamese Air Force MiG-17 with an unguided Zuni rocket as the Skyhawk's only air-to-air victory of the Vietnam war.

From 1956 on, Navy Skyhawks were the first aircraft to be deployed outside of the U.S. armed with the AIM-9 Sidewinder. On strike missions, which was the Skyhawk's normal role, the air-to-air armament was for self defensive purposes.

In the early-to-mid 1960s, standard US Navy A-4B Skyhawk squadrons were assigned to provide daytime fighter protection for ASW aircraft operating from some Essex class US anti-submarine warfare carriers, these aircraft retained their ground- and sea-attack capabilities.

The A-4B model did not have an air-to-air radar, and it required visual identification of targets and guidance from either ships in the fleet or an airborne E-1 Tracer AEW aircraft. Lightweight and safer to land on smaller decks, Skyhawks would later also play a similar role flying from Australian, Argentinean, and Brazilian upgraded World War II surplus light ASW carriers, which were also unable to operate most large modern fighters. Primary air-to-air armament consisted of the internal 20 mm (.79 in) Colt cannons and ability to carry an AIM-9 Sidewinder missile on both underwing hardpoints, later additions of two more underwing hardpoints on some aircraft made for a total capacity of four AAMs.

The first combat loss of an A-4 occurred on 5 August 1964 (fig. 29), when Lieutenant junior grade Alvarez, of VA-144 aboard the USS Constellation, was shot down while attacking enemy torpedo boats in North Vietnam. Alvarez safely ejected after being hit by anti-aircraft artillery (AAA) fire, and became the first US Naval POW of the war; he was released as a POW on 12 February 1973. The last A-4 loss in the Vietnam War occurred on 26 September 1972, when USMC pilot Captain James P. Walsh, USMC of VMA-211, flying from his land base at Bien Hoa Air Base, South Vietnam, was hit by ground fire near An Loc.

An Loc was one of the few remaining hotly contested areas during this time period, and Captain Walsh was providing close air support (CAS) for ground troops in contact (land battle/fire fight) when his A-4 was hit, catching fire, forcing him to eject. Rescue units were sent, but the SAR helicopter was damaged by enemy ground fire, and forced to withdraw. Captain Walsh, after safely ejecting, had landed within NVA (North Vietnamese Army) positions, and had become a POW as soon as his feet had touched the ground. Captain Walsh was the last US Marine to be taken prisoner during the war, and was released as a POW on 12 February 1973.

Although the first A-4Es were flown in Vietnam in early 1965, the A-4Cs continued to be used until late 1970. The Seabees of MCB-10 went ashore on 7 May 1965. On 1 June 1965, the Chu Lai Short Airfield for Tactical Support (SATS) was officially opened with the arrival of eight A-4 Skyhawks from Cubi Point, Philippine Islands. The group landed with the aid of arresting cables, refueled and took off with the aid of JATO, with fuel and bombs to support Marine combat units. The Skyhawks were from Marine Attack Squadron VMA-223 and VMA-311.

On 29 July 1967, the aircraft carrier USS Forrestal was conducting combat operations in the Gulf of Tonkin during the Vietnam War. A Zuni rocket misfired, knocking off an external tank on an A-4. Fuel from the leaking tank caught fire, creating a massive conflagration that burned for hours, killing 134 sailors, and injuring 161. (See 1967 USS Forrestal fire.)

During the war, 362 A-4/TA-4F Skyhawks were lost to all causes. The US Navy lost 271 A-4s, the US Marine Corps lost 81 A-4s and 10 TA-4Fs. A total of 32 A-4s were lost to surface-to-air missiles (SAMs), and one A-4 was lost in aerial combat to a MiG-17 on 25 April 1967.

Fig. 29. A-4C, August 1964.

During the 1982 Falklands War, Argentina deployed 48 Skyhawk warplanes (26 A-4B, 12 A-4C and 10 A-4 aircraft). Armed with unguided bombs and lacking any electronic or missile self-defense, Argentine Air Force Skyhawks sank the Type 42 Destroyer HMS Coventry and the Type 21 Frigate HMS Antelope as well as inflicting heavy damage on several others: the RFA Sir Galahad (1966) (which was subsequently scuttled as a war grave), the Type 42 HMS Glasgow, the Leander Class Frigate HMS Argonaut, the Type 22 Frigate HMS Broadsword, and the RFA Sir Tristram. Argentine Navy A-4Qs, flying from Río Grande, Tierra del Fuego naval air station, also played a role in the bombing attacks against British ships, destroying the Type 21 HMS Ardent (fig. 30).

In all, 22 Skyhawks (10 A-4Ps, nine A-4Cs, and three A-4Qs) were lost to all causes in the six weeks-long war (according to other sources, 23 Skyhawks were lost: 10 A-4Bs, 9 A-4Cs and four A-4Qs).

These losses included eight to British Sea Harriers, seven to ship-launched surface-to-air missiles, four to ground-launched surface-to-air missiles and anti-aircraft fire (including one to "friendly-fire"), and three to crashes.

Fig. 30. A-4C, May 1982, in Argentine.

After the war, Argentine Air Force A-4Ps and A-4Cs survivors were upgraded under the Halcon program with 30 mm (1.2 in) DEFA cannons, air-to-air missiles, and other minor details, and merged into the 5th Air Brigade.

All of these were withdrawn from service in 1999, and they were replaced with 36 of the much improved OA/A-4AR Fightinghawk.

Several TA-4J and A-4E airframes were also delivered under the A-4AR program mainly for spare parts use. In 1983, the United States vetoed the delivery by Israel of 24 A-4Hs for the Argentine Navy as the A-4Q replacement. The A-4Qs were finally retired in 1988.

More recently, Kuwaiti Air Force Skyhawks fought in 1991 during Operation Desert Storm (fig. 31).

When Iraq invaded Kuwait, the available Skyhawks flew attack missions against the advancing Iraqi forces from deserted roads after their bases were overrun. A total of 24 of the 29 A-4KUs that remained in service with Kuwait (from 36 delivered in the 1970s) escaped to Saudi Arabia.

The escaped Skyhawks (along with escaped Mirage F1s) operated as the Free Kuwait Air Force, flying 1,361 sorties during the liberation of Kuwait.

Twenty-three A-4s survived the conflict and the Iraqi invasion, with only one being destroyed in combat. The remaining

Kuwaiti Skyhawks were later sold to Brazil, where they currently serve aboard the aircraft carrier NAe São Paulo.

Fig. 31. A-4C, 1991, Operation Desert Storm in Kuwait.

Fig. 32. This is a modern aircraft A-4, Skyhawk.

In figure 32 it can see a modern aircraft A-4 Skyhawk.

These devices (A-4) have taken the brunt of the fighting over for decades.

Fig. 33. *Airbus A340.*

The Airbus A340 (fig. 33) is a long-range four-engined wide-body commercial passenger jet airliner.

Developed by Airbus Industrie, a consortium of European aerospace companies, which is now fully-owned by EADS, the A340 is manufactured at Toulouse, France. It seats up to 375 passengers in the standard variants and 440 in the stretched −600 series. Depending on the model, it has a range of between 6,700 to 9,000 nautical miles (12,400 to 17,000 km). It is similar in design to the twin-engined A330 with which it was concurrently designed. Its distinguishing features are four high-bypass turbofan engines and three-bogie main landing gear.

Airbus manufactured the A340 in four fuselage lengths. The initial variant, A340-300, entered service in 1993, measured 59.39 metres (194.8 ft), followed by the shorter −200; the stretched A340-600 was a 15.91 metres (52.2 ft) stretch of −200.

This particular variant was developed alongside the shorter A340-500, which would become the longest-ranged commercial airliner until the arrival of the Boeing 777-200LR.

The two initial models were powered by the CFM56-5C, rated at 151 kilonewtons (34,000 lbf), while Rolls-Royce held exclusive powerplant rights to the extended-ranged and heavier −500 and −600 models, through the 267-kilonewton (60,000 lbf) Rolls-Royce Trent 500. Initial A340 versions share the fuselage and wing of the A330 while the −500/-600 models are longer and have larger wings.

Launch customers Lufthansa and Air France placed the A340 into service in March 1993. As of March 2011, 379 orders have been placed (not including private operators), of which 375 have been delivered, with a backlog of four aircraft.

The most common type were the A340-300 model, with 218 aircraft delivered. Lufthansa is the biggest operator of the A340, having acquired 59 aircraft.

The A340 is used on long-haul, trans-oceanic routes due to its immunity from ETOPS; however, with reliability in engines improving, airlines are progressively phasing out the type in favour of more economical twinjets such as the Boeing 777.

The Airbus A380 (fig. 34) is a double-deck, wide-body, four-engine airliner manufactured by the European corporation Airbus, a subsidiary of EADS.

Designed to challenge Boeing's monopoly in the large-aircraft market, the A380, the largest passenger airliner in the world, made its maiden flight on 27 April 2005 from Toulouse, France, and made its first commercial flight on 25 October 2007 from Singapore to Sydney with Singapore Airlines.

The aircraft was known as the Airbus A3XX during much of its development phase, but the nickname Superjumbo has since become associated with it.

The A380's upper deck extends along the entire length of the fuselage, and its width is equivalent to that of a widebody aircraft. This allows for an A380-800's cabin with 5,146 square feet (478.1 m2) of floor space; 49% more floor space than the current next-largest airliner, the Boeing 747-400 with 3,453 square feet (320.8 m2), and provides seating for 525 people in a typical three-class configuration or up to 853 people in all-economy class configurations. The A380-800 has a design range of 15,200 km (8,200 nmi; 9,400 mi), sufficient to fly from New York to Hong Kong for example, and a cruising speed of Mach 0.85 (about 900 km/h or 560 mph at cruising altitude).

As of May 2011, 234 firm orders have been placed, of which 49 had been delivered. The largest operator of the type is Emirates, which has 90 aircraft on order.

Fig. 34. Airbus A380.

In the early 1960s, European air forces began to consider their requirements for the coming decades. One of the results was the emergence of a new generation of jet trainers to replace such aircraft as the Lockheed T-33 and Fouga Magister. The two main rivals in this exercise turned out to be the BAe Hawk and the Franco-German Dassault-Dornier Alpha Jet (fig. 35).

At the outset, the Alpha Jet had a lead, but the BAe Hawk would prove to be the winner in the race. However, the Alpha Jet has been built in good numbers and served with a number of air forces for several decades.

Fig. 35. *Alpha Jet.*

In the early 1960s, the British and French began a collaboration on development of what was supposed to be a supersonic jet trainer/light attack aircraft. The result of this collaboration, the SEPECAT Jaguar, proved to be an excellent aircraft, but its definition had evolved in the interim, and the type emerged as a full-sized strike fighter, with two-seat variants used for operational conversion to the type.

This left the original requirement unfulfilled and so the French began discussions with West Germany for collaboration. A joint specification was produced in 1968. The trainer was now subsonic, supersonic trainers having proven something of a dead end. A joint development and production agreement was signed in July 1969 which indicated that the two nations would buy 200 machines, each assembled in their own country.

The Avro Vulcan (fig. 36), sometimes referred to as the Hawker Siddeley Vulcan, is a delta wing subsonic jet strategic bomber that was operated by the Royal Air Force (RAF) from 1953 until 1984. It was developed by Avro in response to a specification released by the Air Ministry. At the time, both jet engines and delta wings were considered cutting-edge and relatively unexplored; thus, the small-scale Avro 707 was produced to test the principles of the design. In flight, the Vulcan was an agile aircraft for its size.

The Vulcan B.1 was first delivered to the RAF in 1956. In service, the Vulcan was armed with nuclear weapons and was a part of the RAF's V bomber force, the United Kingdom's airborne deterrent against aggression from other powers such as the Soviet Union during the Cold War. In addition to an extensive electronic countermeasures suite, the Vulcan had a small radar cross-section, aiding its deterrent role by evading detection and therefore increasing the likelihood of penetrating Soviet airspace and deploying its weapons load successfully. A second batch of aircraft, the B.2, was produced with new features, including a larger wing and greater fuel capacity, along with more advanced electronics and radar systems.

The B.2s were adapted into several other variants, the B.2A carrying the Blue Steel missile, the B.2 (MRR) for Marine Radar Reconnaissance use, and the K.2 tanker for aerial refuelling. The Vulcan was also used in the secondary role of conventional bombing near the end of its service life in the 1982 Falklands War against Argentina during Operation Black Buck. One example, XH558, was recently restored for use in display flights and commemoration of the employment of the aircraft in the Falklands conflict.

Fig. 36. Avro Vulcan.

The Rockwell (now part of Boeing) B-1 Lancer (fig. 37) is a four-engine, variable-sweep wing strategic bomber used by the United States Air Force (USAF). First envisioned in the 1960s as a supersonic bomber with sufficient range and payload to replace the Boeing B-52 Stratofortress, it developed primarily into a low-level penetrator with long range and supersonic speed capability.

Fig. 37. *B-1 Lancer.*

Designed by Rockwell International, the bomber's development was delayed multiple times over its history, as the theory of strategic balance changed from flexible response to mutually assured destruction and back again. The initial B-1A version was developed in the early 1970s, but its production was canceled and only four prototypes were built. In 1980, the B-1 resurfaced as the B-1B version with the focus on low-level penetration bombing. The B-1B entered service with the USAF in 1986. It began service with the USAF Strategic Air Command as a nuclear bomber.

In the 1990s, the B-1B was converted to conventional bombing use. It was first used in combat during Operation Desert Fox in 1998 and during the NATO action in Kosovo the following year. The B-1B continues to support U.S. and NATO military forces in Afghanistan and Iraq. The Lancer is the supersonic component of the USAF's long-range bomber force, along with the subsonic B-52 and Northrop Grumman B-2 Spirit. The bomber is commonly called the "Bone" (originally from "B-One"). With the retirement of the

General Dynamics/Grumman EF-111A Raven in 1998 and the Grumman F-14 Tomcat in 2006, the B-1B is the U.S. military's only active variable-sweep wing aircraft. The B-1B is expected to continue to serve into the 2020s, when it is to supplemented by the Next Generation Bomber.

The Northrop Grumman B-2 Spirit (fig. 38), (also known as the Stealth Bomber) is an American heavy bomber with low observable stealth technology designed to penetrate dense anti-aircraft defenses and deploy both conventional and nuclear weapons. Because of its considerable capital and operational costs, the project was controversial in the U.S. Congress and among the Joint Chiefs of Staff. During the late 1980s and early 1990s, the Congress slashed initial plans to purchase 132 bombers to 21.

Manufactured by Northrop Grumman, the cost of each aircraft averaged US$737 million in 1997 dollars ($1.01 billion today). Total procurement costs averaged US$929 million per aircraft ($1.27 billion today), which includes spare parts, equipment, retrofitting, and software support. The total program cost, which includes development, engineering and testing, averaged US$2.1 billion per aircraft (in 1997 dollars, $2.87 billion today).

Twenty B-2s are operated by the United States Air Force. Though originally designed in the 1980s for Cold War operations scenarios, B-2s were first used in combat to drop bombs on Serbia during the Kosovo War in 1999, and saw continued use during the wars in Iraq and Afghanistan. One aircraft was lost in 2008 when it crashed just after takeoff; the crew ejected safely. B-2s were also used during the 2011 Libyan uprising.

The bomber has a crew of two and can drop up to 80 500 lb (230 kg)-class JDAM GPS-guided bombs, or 16 2,400 lb (1,100 kg) B83 nuclear bombs in a single pass through extremely dense anti-aircraft defenses.

The B-2 is the only aircraft that can carry large air to surface standoff weapons in a stealth configuration. The program has been the subject of espionage and counter-espionage activity and the B-2 has provided prominent public spectacles at air shows since the 1990s.

Fig. 38. *B-2 Spirit (the Stealth Bomber).*

The Boeing 787 Dreamliner (fig. 39) is a long-range, mid-size wide-body, twin-engine jet airliner developed by Boeing Commercial Airplanes. It seats 210 to 330 passengers, depending on the variant. Boeing states that it is the company's most fuel-efficient airliner and the world's first major airliner to use composite materials for most of its construction. The 787 consumes 20% less fuel than the similarly-sized Boeing 767. Some of its distinguishing features include a four-panel windshield, noise-reducing chevrons on its engine nacelles, and a smoother nose contour.

The aircraft's initial designation was 7E7, prior to its renaming in January 2005. The first 787 was unveiled in a roll-out ceremony on July 8, 2007, at Boeing's Everett assembly factory, by which time it had become the fastest-selling wide-body airliner in history with 677 orders. By March 2011, 835 Boeing 787s had been ordered by 56 customers. As of 2011, launch customer All Nippon Airways has the largest number of 787s on order.

Fig. 39. *Boeing 787 Dreamliner.*

The 787 development and production has involved a large-scale collaboration with numerous suppliers around the globe. It is being assembled at the Boeing Everett Factory in Everett, Washington. Aircraft will also be assembled at a new factory in North

Charleston, South Carolina. Both sites will deliver 787s to airline customers. Originally planned to enter service in May 2008, the project has suffered from repeated delays and is now more than three years behind schedule. The airliner's maiden flight took place on December 15, 2009, and it is currently undergoing flight testing with a goal of receiving certification in mid-2011 and entering service with All Nippon Airways in the third quarter of 2011.

The Aérospatiale-BAC Concorde (fig. 40) is a turbojet-powered supersonic passenger airliner, a supersonic transport (SST). It was a product of an Anglo-French government treaty, combining the manufacturing efforts of Aérospatiale and the British Aircraft Corporation. First flown in 1969, Concorde entered service in 1976 and continued commercial flights for 27 years.

Among other destinations, Concorde flew regular transatlantic flights from London Heathrow (British Airways) and Paris-Charles de Gaulle Airport (Air France) to New York JFK, profitably flying these routes at record speeds, in less than half the time of other airliners.

Fig. 40. Concorde (is a turbojet-powered supersonic passenger).

With only 20 aircraft built, their development represented a substantial economic loss, in addition to which Air France and British Airways were subsidised by their governments to buy them. As a result of the type's only crash on 25 July 2000 and other factors, its retirement flight was on 26 November 2003.

Concorde's name reflects the development agreement between the United Kingdom and France. In the UK, any or all of the type—unusual for an aircraft—are known simply as "Concorde". The aircraft is regarded by many as an aviation icon.

The Lockheed F-117 Nighthawk (fig. 41) is a single-seat, twin-engine stealth ground-attack aircraft formerly operated by the United States Air Force (USAF). The F-117A's first flight was in 1981, and it achieved initial operating capability status in October 1983. The F-117A was "acknowledged" and revealed to the world in November 1988.

***Fig. 41.** The Lockheed F-117 Nighthawk.*

A product of Lockheed Skunk Works and a development of the Have Blue technology demonstrator, it became the first operational aircraft initially designed around stealth technology. The F-117A was widely publicized during the Persian Gulf War of 1991. It was commonly called the "Stealth Fighter" although it was a ground-attack aircraft, making its F-designation misleading.

The Air Force retired the F-117 on 22 April 2008, primarily due to the fielding of the F-22 Raptor and the impending introduction of the F-35 Lightning II. Sixty-four F-117s were built, 59 of which were production versions with 5 demonstrators/prototypes.

Fig. 42. *The Grumman F-14 Tomcat.*

The Grumman F-14 Tomcat (fig. 42) is a supersonic, twin-engine, two-seat, variable-sweep wing fighter aircraft. The Tomcat was developed for United States Navy's Naval Fighter Experimental (VFX) program following the collapse of the F-111B project. The F-14 was the first of the American teen-series fighters which were designed incorporating the experience of air combat against MiGs during the Vietnam War.

The F-14 first flew in December 1970. It first deployed in 1974 with the U.S. Navy aboard USS Enterprise (CVN-65), replacing the McDonnell Douglas F-4 Phantom II. The F-14 served as the U.S. Navy's primary maritime air superiority fighter, fleet defense interceptor and tactical reconnaissance platform. In the 1990s it added the Low Altitude Navigation and Targeting Infrared for Night (LANTIRN) pod system and began performing precision strike missions. The F-14 was retired from the active U.S. Navy fleet on 22 September 2006, having been replaced by the Boeing F/A-18E/F Super Hornet. As of 2009, the F-14 was only in service with the Islamic Republic of Iran Air Force, having been exported to Iran in 1976 when the US had amicable diplomatic relations with the nation.

The General Dynamics F-16 Fighting Falcon (fig. 43) is a multirole jet fighter aircraft originally developed by General Dynamics for the United States Air Force (USAF). Designed as a lightweight day fighter, it evolved into a successful all-weather multirole aircraft. Over 4,400 aircraft have been built since production was approved in 1976. Though no longer being purchased by the U.S. Air Force, improved versions are still being built for export customers. In 1993, General Dynamics sold its aircraft manufacturing business to the Lockheed Corporation, which in turn became part of Lockheed Martin after a 1995 merger with Martin Marietta.

The Fighting Falcon is a dogfighter with numerous innovations including a frameless bubble canopy for better visibility, side-mounted control stick to ease control while maneuvering, a seat reclined 30 degrees to reduce the effect of g-forces on the pilot, and the first use of a relaxed static stability/fly-by-wire flight control system that makes it a highly nimble aircraft. The F-16 has an internal M61 Vulcan cannon and has 11 hardpoints for mounting weapons, and other mission equipment. Although the F-16's official name is "Fighting Falcon", it is known to its pilots as the "Viper", due to it resembling a viper snake and after the Battlestar Galactica Colonial Viper starfighter.

In addition to USAF active, reserve, and air national guard units, the aircraft is used by the USAF aerial demonstration team, the U.S. Air Force Thunderbirds, and as an adversary/aggressor aircraft by the United States Navy. The F-16 has also been procured to serve in the air forces of 25 other nations.

Fig. 43. *The F-16 Fighting Falcon over Iraq.*

The Boeing F/A-18E/F Super Hornet (fig. 44) is a twin-engine carrier-based multirole fighter aircraft. The F/A-18E single-seat variant and F/A-18F tandem-seat variant are larger and more advanced derivatives of the F/A-18C and D Hornet. The Super Hornet has an internal 20 mm gun and can carry air-to-air missiles and air-to-surface weapons. Additional fuel can be carried with up to five external fuel tanks and the aircraft can be configured as an airborne tanker by adding an external air refueling system.

Fig. 44. The F-18 Super Hornet.

Designed and initially produced by McDonnell Douglas, the Super Hornet first flew in 1995. Full-rate production began in September 1997, after the merger of McDonnell Douglas and Boeing the previous month. The Super Hornet entered service with the United States Navy in 1999, replacing the Grumman F-14 Tomcat since 2006, and serves alongside the original Hornet. The Royal Australian Air Force (RAAF), which has operated the F/A-18A as its main fighter since 1984, ordered the F/A-18F in 2007 to replace its aging F-111 fleet. RAAF Super Hornets entered service in December 2010.

The Lockheed Martin/Boeing F-22 Raptor (fig. 45) is a single-seat, twin-engine fifth-generation supermaneuverable fighter aircraft that uses stealth technology. It was designed primarily as an air superiority fighter, but has additional capabilities that include ground attack, electronic warfare, and signals intelligence roles. Lockheed Martin Aeronautics is the prime contractor and is responsible for the majority of the airframe, weapon systems and final assembly of the F-

22. Program partner Boeing Defense, Space & Security provides the wings, aft fuselage, avionics integration, and all of the pilot and maintenance training systems.

Fig. 45. *The Lockheed Martin/Boeing F-22 Raptor.*

The aircraft was variously designated F-22 and F/A-22 during the years prior to formally entering USAF service in December 2005 as the F-22A.

Despite a protracted and costly development period, the United States Air Force considers the F-22 a critical component for the future of US tactical air power, and claims that the aircraft is unmatched by any known or projected fighter, while Lockheed Martin claims that the Raptor's combination of stealth, speed, agility, precision and situational awareness, combined with air-to-air and air-to-ground combat capabilities, makes it the best overall fighter in the world today.

Air Chief Marshal Angus Houston, Chief of the Australian Defence Force, said in 2004 that the "F-22 will be the most outstanding fighter plane ever built."

The high cost of the aircraft, a lack of clear air-to-air combat missions because of delays in the Russian and Chinese fifth generation fighter programs, a US ban on Raptor exports, and the ongoing development of the supposedly cheaper and more versatile F-35 resulted in calls to end F-22 production.

In April 2009 the US Department of Defense proposed to cease placing new orders, subject to Congressional approval, for a final procurement tally of 187 Raptors.

The National Defense Authorization Act for Fiscal Year 2010 was signed into law in October 2009 without funding for further F-22 production.

The Lockheed Martin F-35 Lightning II (fig. 46) is a family of single-seat, single-engine, fifth generation multirole fighters under development to perform ground attack, reconnaissance, and air defense missions with stealth capability.

The F-35 has three main models; one is a conventional takeoff and landing variant, the second is a short take off and vertical-landing variant, and the third is a carrier-based variant.

The F-35 is descended from the X-35, the product of the Joint Strike Fighter (JSF) program.

JSF development is being principally funded by the United States, with the United Kingdom and other partner governments providing additional funding.

It is being designed and built by an aerospace industry team led by Lockheed Martin.

The F-35 took its first flight on 15 December 2006.

The United States intends to buy a total of 2,443 aircraft for an estimated US$323 billion, making it the most expensive defense program ever.

The United States Air Force (USAF) budget data in 2010, along with other sources, projects the F-35 to have a flyaway cost from US$89 million to US$200 million over the planned production of F-35s. Cost estimates have risen to $382 billion for 2,443 aircraft, at an average of $156 million each.

The rising program cost estimates have cast doubt on the actual number to be produced for the U.S. In January 2011, the F-35B variant was placed on "probation" for two years because of development issues.

In February 2011, the Pentagon put a price of $207.6 million for each of the 32 aircraft to be acquired in FY2012, rising to $304.15 million ($9,732.8/32) if its share of RDT&E spending is included.

Fig. 46. *The Lockheed Martin F-35 Lightning II.*

Beginning with the Vietnam War doubts began to be raised about the ability of the 700+ KC-135 fleet to meet the needs of the United States' global commitments.

The aerial refueling fleet was deployed to Southeast Asia in support of tactical aircraft and strategic bombers, while maintaining the US-based support of the nuclear bomber fleet.

The United States Air Force as a result sought an air-to-air tanker with a greater capability than the KC-135. In 1972 two DC-10s were flown in trials at Edwards Air Force Base, simulating air refuelings to check for possible wake issues. Boeing performed similar tests with a 747.

The 1973 Yom Kippur War and the US Operation Nickel Grass demonstrated the necessity of adequate air-refueling capabilities. Denied landing rights in Europe, USAF C-5 Galaxies were forced to carry a fraction of their maximum payload on direct flights from the continental United States to Israel.

As a result C-5 crews were soon trained in aerial refueling and the U.S. Department of Defense concluded that a more advanced tanker was needed.

In 1975, under the Advance Tanker Cargo Aircraft program, four aircraft were evaluated: the C-5 itself, the Boeing 747, the McDonnell Douglas DC-10, and the Lockheed L-1011. The U.S Air Force selected McDonnell Douglas's DC-10 over Boeing's 747 in December 1977.

The design for the KC-10 involved only modifications from the DC-10-30CF design. The major changes were the addition of a boom control station in the rear of the fuselage and extra fuel tanks below the main deck. The KC-10 has both a centerline refueling boom and a drogue/hose system on the right side of the rear fuselage. Other changes from the DC-10-30CF include the removal of most cargo doors and windows.

The KC-10 first flew on 12 July 1980. Early aircraft featured a paint scheme with light gray on the airplane's belly and white on the upper portion with blue around the cockpit. A gray-green camouflage scheme was used on later tankers. Aircraft have since been switched to a medium gray color. The KC-10 boom operator is located in the rear of the airplane with wide window for monitoring refueling. The operator controls refueling operations through a digital, fly-by wire system.

The final 20 KC-10s produced included wing-mounted pods for added refueling locations. In addition to the USAF refueling boom, the KC-10's hose and drogue system allows refueling of U.S. Navy, Marine Corps, and most NATO allied aircraft. This gives the KC-10 the ability to refuel US and other NATO aircraft, all in one mission.

A need for new transport aircraft for the Royal Netherlands Air Force was first identified in 1984. In 1991 four categories of transport requirements were established. Category A required a large cargo aircraft with a range of at least 4500 km and the capability to refuel F-16s. In 1992, 2 DC-10-30CFs were acquired from Martinair in a buy/leaseback contract. When one of the bought aircraft was lost in the Martinair Flight 495 crash, a third aircraft was bought from Martinair.

The conversion was handled via the United States foreign military sales program, which in turn contracted McDonnell Douglas, the designer of both the DC-10 and the KC-10 tanker. Costs for the conversion were initially estimated at $89.5 million (FY 1994). The aircraft was to be equipped with both a boom and a probe and drogue system. However, because McDonnell Douglas did not have any experience with the requested Remote Aerial Refueling Operator (RARO) system, and because the third aircraft differed from the original two, the program could not be completed at budget. By omitting the probe and drogue system and a fixed partition wall between the cargo and passenger, the cost could be limited at $96 million. To make up for the cost increase McDonnell Douglas hired Dutch companies to do part of the work. The actual converting of the aircraft for instance was done by KLM. Conversion of the aircraft was done from October 1994 to September 1995 for the first aircraft and from February to December 1995 for the second. This was much longer than planned, mostly because McDonnell Douglas did not deliver the parts in time. This would have again increased the cost, but in the contract for the AH-64 Apaches which the Royal Netherlands Air Force also bought from McDonnell Douglas, the price was agreed to be kept at $96 million.

In 2005 Fokker Services (NL) was awarded with a contract to update the avionics on the two KDC-10 and one DC-10 aircraft of the Royal Netherlands Air Force to a common standard. The aircraft will be updated with digital cockpits, Link 16 and satellite communications. This design is now offered by Boeing to the US Air Force to update their KC-10s.

The KC-10 was delivered to the USAF Strategic Air Command (SAC), then in control of air refueling assets from 1981 to 1987. SAC had KC-10 Extenders (fig. 55) in service from 1981 until 1992, when they were re-assigned to the newly established Air Mobility Command. In the air-to-air refueling (AAR) role, the KC-10s have been operated largely in the strategic refueling of large number of tactical aircraft on ferry flights and the refueling of other strategic transport aircraft. Conversely, the KC-135 fleet has operated largely in the in-theater tactical role.

There are 59 KC-10 Extenders (fig. 47) in service with the United States Air Force (USAF) as of 2010. The KC-10 has a significantly larger fuel capacity than the Air Force's other main tanker, the KC-135. with over 440 in service in 2009. The USAF's

KC-10s are stationed primarily at Travis AFB, California and McGuire AFB, New Jersey.

When faced with refusals of basing and overflight rights from continental European countries during Operation El Dorado Canyon, the U.S. was forced to use the UK-based F-111s in the 1986 air-strikes against Libya. The KC-10s allowed 29 F-111s to reach their targets. The KC-10 fleet also facilitated the deployment of tactical, strategic, and transport aircraft to Saudi Arabia during Operation Desert Shield. In recent times, USAF KC-10s have seen use in supporting military operations in Iraq and Afghanistan through 2009. In addition to offering in-flight refueling capabilities, USAF KC-10s also undertake cargo and personnel transport operations.

In an attempt to modernize the platform, the USAF has awarded Boeing a US$216 million contract to upgrade its fleet of 59 aircraft with new communication, navigation, surveillance and air traffic management (CNS/ATM) system. Boeing claims that this will allow the aircraft to fly in civil airspace after 2015 as new ICAO and FAA standards take effect.

Fig. 47. The KC-10 Extenders.

The Lockheed C-5 Galaxy (fig. 48) is a large military transport aircraft built by Lockheed. It was designed to provide strategic heavy airlift over intercontinental distances and to carry outsize and oversize cargo. The C-5 Galaxy has been operated by the United States Air Force (USAF) since 1969 and is among the largest military aircraft in the world. The C-5M Super Galaxy is an upgraded version with new engines and modernized avionics designed to extend its service life beyond 2040.

Fig. 48. The Lockheed C-5 Galaxy.

In 1961, several aircraft companies began studying heavy jet transport designs that would replace the Douglas C-133 Cargomaster transport and complement Lockheed C-141 Starlifters. In addition to higher overall performance, the United States Army wanted a transport with a larger cargo bay than the C-141, whose interior was too small to carry a variety of their outsized equipment. These studies led to the "CX-4" design concept, but in 1962 the proposed six-engine design was rejected, because it was not viewed as a significant advance over the C-141.

By late 1963, the next conceptual design was named CX-X. It was equipped with four engines, instead of six engines in the earlier CX-4 concept. The CX-X had a gross weight of 550,000 pounds (249,000 kg), a maximum payload of 180,000 lb (81,600 kg) and a speed of Mach 0.75 (500 mph/805 km/h). The cargo compartment was 17.2 ft (5.24 m) wide by 13.5 feet (4.11 m) high and 100 ft (30.5 m) long with front and rear access doors. To provide required power and range with only four engines required a new engine with dramatically improved fuel efficiency.

The criteria were finalized and an official Request for Proposal was sent out in April 1964 for the "Heavy Logistics System" (CX-HLS) (previously CX-X). In May 1964, proposals for aircraft were received from Boeing, Douglas, General Dynamics, Lockheed, and Martin Marietta. Proposals for engines were received from General Electric, Curtiss-Wright, and Pratt & Whitney. After a downselect, Boeing, Douglas and Lockheed were given one-year study contracts for the airframe, along with General Electric and Pratt & Whitney for the engines.

All three of the designs shared a number of features. In particular, all three placed the cockpit well above the cargo area to allow for cargo loading through a nose door. The Boeing and Douglas designs used a pod on the top of the fuselage containing the cockpit, while the Lockheed design extended the cockpit profile down the length of the fuselage, giving it an egg-shaped cross section. All of the designs featured swept wings and front and rear cargo doors allowing simultaneous loading and unloading. Lockheed's design featured a T-tail, while the designs by Boeing and Douglas had conventional tails.

The Air Force considered Boeing's design better than the Lockheed design, although Lockheed's proposal was the lowest total cost bid. Lockheed was selected the winner in September 1965, then awarded a contract in December 1965. General Electric's engine design was selected in August 1965 for the new transport; the Pratt & Whitney engine design was developed and later used on the Boeing 747.

The first C-5A Galaxy (number 66-8303) was rolled out of the manufacturing plant in Marietta, Georgia on 2 March 1968. On 20 June 1968, Lockheed-Georgia Co. began flight testing its new Galaxy C-5A heavy transport with the aircraft's first flight taking to the air under the call-sign "eight-three-oh-three heavy" (8303H). Cost overruns and technical problems of the C-5A were the subject of a congressional investigation in 1968 and 1969. The Lockheed C-5 program holds the dubious distinction of being the first program to produce a one billion dollar overrun.

Upon completion of testing the first C-5A was transferred to the Transitional Training Unit at Altus Air Force Base, OK, in December 1969. Lockheed then delivered the first operational Galaxy to the 437th Airlift Wing, Charleston Air Force Base, SC, in June 1970. Due to the higher than expected development costs, there were calls within the military as early as in June 1970 for the government to split the losses that Lockheed were experiencing. Production was nearly brought to a halt in 1971 due to Lockheed going through financial difficulties, brought on in part due to the C-5 Galaxy but also by the civilian jet liner, the Lockheed L-1011. The U.S. government loaned Lockheed money so they could keep running.

In 1969 Henry Durham raised concerns about the C-5 production process with his employer, Lockheed. He was transferred and subjected to abuse until he resigned from the company. The GAO substantiated some of his charges against Lockheed and the American Ethical Union honored him with the Elliott-Black Award.

In the early 1970s, NASA considered the C-5 for the Shuttle Carrier Aircraft role, to transport the Space Shuttle to Kennedy Space Center. However, they rejected it in favor of the Boeing 747, in part due to the 747's low-wing design. In contrast, the Soviet Union chose to transport its shuttles using the high-winged An-225, which derives from the An-124, which is similar in design and function to the C-5.

During static and fatigue testing cracks were noticed in the wings of several aircraft, and as a consequence the C-5A fleet was restricted to 80% of maximum design loads. To reduce wing loading, load alleviation systems were added to the aircraft. By 1980, payloads were restricted to as low as 50,000 lb (23,000 kg) for general cargo during peacetime operations. To restore full payload capability and service life, a $1.5 billion program to re-wing the 76 completed C-5As began in 1976. After design and testing of the new wing design, the C-5As received their new wings from 1980 to 1987. During 1976, numerous cracks were also found in the fuselage along the upper fuselage on the centerline, aft of the refueling port, extending back to the wing. The cracks required a redesign to the hydraulic system for the visor, the front cargo entry point.

In 1974, Iran, then a monarchy which maintained good relations with the United States, offered $160 million to restart the production of the C-5 to enable them to make their own procurements of the Galaxy; in a similar climate as to their acquisition of F-14 Tomcat fighters. However no C-5 aircraft were ever ordered by Iran, as the prospect was firmly halted by the Iranian Revolution in 1979.

As part of President Ronald Reagan's military planning, a new version of the C-5, the C-5B, was approved by Congress for purchase in July 1982. The first C-5B was delivered to Altus Air Force Base in January 1986. In April 1989, the last of 50 C-5B aircraft was added to the 77 C-5As in the Air Force's airlift force structure. The C-5B includes all C-5A improvements and numerous additional system modifications to improve reliability and maintainability.

In 1998, the Avionics Modernization Program (AMP) began upgrading the C-5's avionics to include a glass cockpit, navigation equipment, and a new autopilot system. Another part of the C-5 modernization effort is the Reliability Enhancement and Re-engining Program (RERP). The program will mainly replace the engines with newer, more powerful ones. Three C-5s are to undergo RERP as a test with full production planned to begin in May 2008.

The C-5 is a large high-wing cargo aircraft. It has a distinctive high T-tail, 25 degree wing sweep, and four TF39 turbofan engines mounted on pylons beneath the wings. The C-5 is similar in layout to its smaller predecessor, the C-141 Starlifter. The C-5 has 12 internal wing tanks and is equipped for aerial refueling. It has both nose and aft doors for "drive-through" loading and unloading of cargo.

The C-5 features a cargo compartment 121 ft (37 m) long, 13.5 ft (4.1 m) high, and 19 ft (5.8 m) wide, or just over 31,000 cu ft (880 m3). The compartment can accommodate up to 36 463L master pallets or a mix of palletized cargo and vehicles. The cargo hold of the C-5 is actually a foot longer than the length of the first powered flight by the Wright Brothers' Flyer at Kitty Hawk. The nose and aft doors open the full width and height of the cargo compartment to permit faster and easier loading. Ramps are full width at each end for loading double rows of vehicles.

It has an upper deck seating area for 73 passengers. The passengers face the rear of the aircraft, rather than forward. Its takeoff and landing distances, at maximum gross weight, are 8,300 ft (2,500 m) and 4,900 ft (1,500 m) respectively. Its high flotation main landing gear has 28 wheels to share the weight. The rear main landing gear is steerable for a smaller turning radius and it rotates 90 degrees horizontally before it is retracted after takeoff. The "kneeling" landing gear system permits lowering of the parked aircraft so the cargo floor is at truck-bed height to facilitate vehicle loading and unloading.

The C-5 has a Malfunction Detection Analysis and Recording (MADAR) system, which records and analyzes information and detects malfunctions in more than 800 test points. The C-5 requires an average of 16 hours of maintenance for each flight hour based on 1996 data. The Galaxy has a Low Pressure Pneumatic System (LPPS) that utilizes a turbo compressor driven by bleed air to provide 150 psi pressure for inflating aircraft's tires on the ground. The LPPS is no longer used on C-5s.

The Galaxy is capable of carrying nearly every type of the Army's combat equipment, including bulky items such as the 74 short tons (67 t) armored vehicle launched bridge (AVLB), from the United States to any location on the globe.

One of the unique features was the crosswind landing system that allows the landing gear to be offset up to 20 degrees either side of centerline. When the main landing gear was down (MLG) all the other 28 wheels would be slaved to the MLG and driven by hydraulic actuators to the same offset.

The first C-5A was delivered to the USAF on 17 December 1969. Wings were built up in the early 1970s at Altus AFB, Oklahoma, Charleston AFB, Dover AFB, Delaware, and Travis AFB, California. 9 July 1970 marked the C-5's first mission in Southeast Asia during the Vietnam War. Through the rest of the war, C-5s were used to transport equipment and troops, including Army tanks and various smaller aircraft. C-5s have also been used to deliver support and reinforce various U.S. allies over the years, critically delivered weapons and supplies to Israel as part of Operation Nickel Grass in 1973, in which the aircraft performed to such a high degree that the Pentagon considered further purchases. The C-5 was also made available to support British-led peacekeeping efforts in Zimbabwe in 1979.

The C-5 is the largest aircraft to ever operate in the Antarctic. Williams Field near McMurdo Station is capable of handling C-5 aircraft and the first C-5 landed there in 1989. The C-5 Galaxy was a core part of the extensive airlift operations supplying troops involved in the First Gulf War, and in delivering relief aid to Rwanda in 1994.

The wings on the C-5As were replaced during the 1980s to restore full design capability. The U.S. Air Force took delivery of the first C-5B on 28 December 1985 and the final one in April 1989. The reliability of the C-5 fleet has been a continued issue throughout its lifetime, however the C-5M upgrade program seeks in part to address this issue.

In response to Air Force motions towards the retirement of the C-5 Galaxy, Congress implemented legislation that placed set limits upon retirement plans for C-5A models in 2003. By 2005, 14 C-5As were retired. One was sent to the Warner Robins Air Logistics Center (WR-ALC) for tear down and inspection to evaluate structural integrity and estimate the remaining life for the fleet. Thirteen C-5As were sent to the Air Force's Aerospace Maintenance and Regeneration Group (AMARG) for inspection of levels of corrosion and fatigue.

In 2007 the Air Force requested information on the Airbus A380 freighter for possible use as a military transport to supplement the U.S. strategic airlift fleet.

The U.S. Air Force began to receive refitted C-5M aircraft in December 2008; full production of C-5Ms began in the summer of 2009. In 2009, the Congressional ban on the retirement of C-5s was overturned. The Air Force seeks to retire one C-5A for each 10 new C-17s ordered.

The C-5A is the original version of the C-5. From 1969 to 1973, 81 C-5As were delivered to U.S. Air Forces bases. Due to cracks found in the wings in the mid-1970s, the cargo weight was restricted.

To restore the plane's full capability, the wing structure was redesigned.

A program to install new strengthened wings on 77 C-5As was conducted from 1981 to 1987.

The redesigned wing made use of a new aluminum alloy that did not exist during the original production.

The C-5B is an improved version of the C-5A. It incorporated all modifications and improvements made to the C-5A with improved wings, upgraded TF-39-GE-1C turbofan engines and updated avionics.

From 1986 to 1989, 50 of the new variant were delivered to the U.S. Air Force.

The C-5C is a specially modified variant for transporting large cargo. Two C-5s (68-0213 and 68-0216) were modified to have a larger internal cargo capacity to accommodate large payloads, such as satellites for use by NASA.

The major modifications were the removal of the rear passenger compartment floor, splitting the rear cargo door in the middle, and installing a new movable aft bulkhead further to the rear. Modifications also included adding a second inlet for ground power, which can feed any power-dependent equipment that may form part of the cargo.

The two C-5Cs are operated by U.S. Air Force crews on the behalf of NASA, and are stationed at Travis AFB, California. 68-0216 completed the Avionics Modernization Program in January 2007.

Based on a recent study showing 80% of the C-5 airframe service life remaining, AMC began an aggressive program to modernize all remaining C-5Bs and C-5Cs and many of the C-5As. The C-5 Avionics Modernization Program (AMP) began in 1998 and includes upgrading avionics to Global Air Traffic Management compliance, improving communications, new flat panel displays, improving navigation and safety equipment, and installing a new autopilot system. The first flight of the first modified C-5 with AMP (85-0004) occurred on 21 December 2002.

Another part of the plan is a comprehensive Reliability Enhancement and Re-engining Program (RERP), which includes new General Electric CF6-80C2 engines, pylons and auxiliary power units, with upgrades to aircraft skin and frame, landing gear, cockpit and the pressurization system. The CF6 engine produces 22% more thrust (for 50,000 lbf/220 kN total from each engine) than existing C-5 engines, which will result in a 30% shorter takeoff roll, a 38% higher climb rate to initial altitude, a significantly increased cargo load, and a longer range between refueling. The C-5s that complete these upgrades are designated C-5M Super Galaxy.

The C-5 AMP and RERP modernization programs plan to raise mission-capable rate to a minimum goal of 75%. Over the next 40 years, the U.S. Air Force estimates the C-5M will save over $20 billion. The first C-5M conversion was completed on 16 May 2006, and performed its first flight on 19 June 2006. C-5Ms have been in flight testing out of Dobbins Air Reserve Base since June 2006. Test aircraft include a distinctively colored nose boom to acquire flight data.

The USAF decided to convert remaining C-5Bs and C-5Cs into C-5Ms with avionics upgrades and re-engining in February 2008. The C-5As will receive only the avionics upgrades. The three test C-5Ms successfully completed developmental flight testing in August 2008. The test aircraft will begin Operational Test and Evaluation in September 2009. The RERP upgrade program is to be completed in 2016.

Lockheed Martin announced that a C-5M test flight on 13 September 2009, set 41 new records. The flight's data have been submitted to the National Aeronautic Association for formal acceptance. The C-5M carried a payload of 176,610 lb (80,110 kg) to over 41,100 ft (12,500 m) in 23 minutes, 59 seconds. The flight set 33 time to climb records at various payload classes, and broke the world record for greatest payload to 6,562 feet (2,000 meters). The aircraft used for this flight had a takeoff weight of 649,680 lb (294,690 kg), which included payload, fuel and crew.

A total of 52 C-5s are contracted to be modernized, consisting of 49 B-, two C- and one A-model aircraft through the Reliability Enhancement and Re-Engining Program (RERP). Over 70 changes and upgrades are incorporated in the program, including newer, quieter, General Electric engines which increase reliability and make the Super Galaxy 10 percent more fuel efficient than legacy C-5s. Five C-5M Super Galaxies have been produced.

Lockheed also planned a civilian version of the C-5 Galaxy, the L-500, the company designation also used for the C-5 itself. Both passenger and cargo versions of the L-500 were designed.

The all-passenger version would have been able to carry up to 1,000 travelers, while the all-cargo version was predicted to be able to carry typical C-5 volume for as little as 2 cents per ton-mile (in 1967 dollars).

Although some interest was expressed by carriers, no orders were placed for either L-500 version, due to operational costs caused by low fuel efficiency, a significant concern for a profit-making carrier, even before the oil crises of the 1970s, keen competition from

Boeing's 747, and high costs incurred by Lockheed in developing the C-5 and later, the L-1011 which led to the governmental rescue of the company.

There have been five C-5 Galaxy aircraft lost in crashes along with two class-A losses resulting from ground fire and one loss resulting from damage sustained on the ground. There have been at least two other C-5 crashes that resulted in major airframe damage, but the aircraft were repaired and returned to service.

On 27 May 1970, C-5A serial number 67-0172 was destroyed during a ground fire at Palmdale, California after an Air Turbine Motor (ATM) started backwards and quickly overheated, setting the hydraulic system on fire and quickly consuming the aircraft.

The engines were not running at the time of the fire. Five crew escaped, but seven firefighters suffered minor injuries fighting the blaze.

On 17 October 1970, C-5A S/N 66-8303 was destroyed during a ground fire at Marietta, Georgia. The fire started during maintenance in one of the aircraft's 12 fuel cells. One worker was killed and another injured. This was the first C-5 aircraft produced.

On 27 September 1974, C-5A 68-0227 crashed after overrunning the runway at Clinton, Oklahoma Municipal Airport during an emergency landing following a serious landing gear fire. The crew mistakenly aligned the aircraft for the visual approach into the wrong airport, landing at Clinton Municipal Airport, which has a 4,400 ft (1,300 m) runway—instead of Clinton-Sherman airfield, which has a 13,500 ft (4,100 m) runway. This was the first operational loss of a C-5 Galaxy.

On 4 April 1975, C-5A 68-0218 crashed while carrying orphans out of Vietnam (Operation Baby Lift). This crash, known as Tan Son Nhut C-5 accident, is one of the most well known C-5 accidents to date. The crash occurred while trying to make an emergency landing at Tan Son Nhut Air Base Saigon, following a door lock failure in flight. 144 adults and children (including 76 babies) were killed out of the 305 aboard (243 children, 44 escorts, 16 crewmen and two flight nurses). Use of the C-5 was heavily restricted for several months due to this high profile incident.

On 31 July 1983, C-5A 70-0446 crashed on landing at Shemya, Alaska. The C-5 approached below the glide slope, hit an embankment short of the runway and bounced back into the air before coming to rest on the runway. Structural damage was extensive and the two aft main landing gear bogies were sheared from the airplane. There were no fatalities. A joint USAF/Lockheed team made repairs enabling a one-time ferry flight from Shemya to the Lockheed plant in Marietta, Georgia. There, the airplane was quickly christened Phoenix II and permanent repair efforts got underway. In addition to the structural repairs, Phoenix II also received an improved landing gear system (common to the then-new C-5B), wing modification, and a color weather radar upgrade. The airplane was returned to service, and was transferred to the Texas Air National Guard.

In July 1985, C-5A 68-0216 landed wheels (gear) up at Travis Air Force Base, California. There were no injuries. The accident occurred while the crew was performing touch-and-go landings, and did not lower the landing gear during the final approach of the day. The aircraft received significant damage to the lower fuselage and main landing gear pods. The C-5A was later flown to Marietta for repairs. While there, the aircraft was selected to be the first C-5A converted to the C-5C configuration.

On 29 August 1990, C-5A 68-0228 crashed following an engine failure shortly after take-off. The aircraft took off from Ramstein Air Base in Germany in support of Operation Desert Shield. It was flown by a nine-member reserve crew from the 68th Airlift Squadron, 433rd Airlift Wing based at Kelly AFB, Texas. As the aircraft started to climb off the runway, one of the thrust reversers suddenly deployed. This resulted in loss of control of the aircraft and the subsequent crash. Of the 17 people on board, only four survived the crash. All four were in the rear troop compartment. The sole crew member to survive, Staff Sgt. Lorenzo Galvan, Jr., was awarded the Airman's Medal for his actions in evacuating the survivors from the wreckage.

On 3 April 2006, C-5B 84-0059 crashed after an in-flight emergency involving an indication that a thrust reverser was not locked. The C-5B assigned to the 436th Airlift Wing and flown by a reserve crew from the 709th Airlift Squadron, 512th Airlift Wing crashed about 2,000 ft (610 m) short of runway 32, while attempting a heavyweight emergency landing at Dover Air Force Base, Delaware. The airplane, carrying 17 people, had taken off from Dover about 21 minutes earlier and reported an in-flight emergency (number two engine thrust reverser not locked indication) 10 minutes into the flight.

All 17 aboard survived, but two received serious injures.

The Air Force's accident investigation concluded the crash was a result of human error, most notably that the crew kept one of the functioning engines in flight idle while manipulating the throttle of the (dead) number two engine as if it was still running, while having the number three engine at idle, an error that was further amplified by the crew's decision to use a high flap setting that increased drag beyond normal two engine performance capabilities.

The forward fuselage will be converted into a C-5 AMP avionics test bed, and the rest of the airframe has been scrapped.

The Saab 37 Viggen (Fig. 49), (English: Thunderbolt) was a Swedish single-seat, single-engine, short-medium range fighter and attack aircraft, manufactured between 1970 and 1990. Several variants were produced to perform the roles of all-weather fighter-interceptor, ground-attack and photo-reconnaissance, as well as a two-seat trainer.

Fig. 49. The SAAB JA 37 Viggen.

The Viggen was initially developed as a replacement for the Saab 32 Lansen in the attack role and later the Saab 35 Draken as a fighter. The first studies were carried out between 1952 and 1957 involving the Finnish aircraft designer Aarne Lakomaa. Several different concepts were studied involving both single- and twin-engines and also with separate lift engines, both simple and double delta wings and also with canard wings. Even VTOL designs were considered.

The aim was to produce a robust aircraft with good short-runway performance that could be operated from numerous specially prepared roads and highways to reduce the vulnerability to attack in the event of war. Other requirements included supersonic ability at low level, Mach 2 performance at altitude, and the ability to make short landings at low angles of attack (to avoid damaging improvised runways). The aircraft was also designed from the beginning to be easy to repair and service, even for personnel without much training.

To meet these design goals, Saab selected a radical configuration: a conventional delta wing with small, high-set canard wings. Canards have since become common in fighter aircraft, notably with the Eurofighter Typhoon, Dassault Rafale, Saab JAS 39 Gripen and the IAI Kfir, but mainly for agility reasons rather than STOL capabilities. The final proposal was presented and accepted on 28 September 1962. Construction started in 1964, with a first prototype maiden flight on 8 February 1967.

In 1960, the U.S. National Security Council, led by President Eisenhower, formulated a military security guarantee for Sweden. The U.S. promised to help the Swedish militarily in the event of a Soviet attack against Sweden; both countries signed a military-technology agreement. In what was known as the "37-annex", Sweden was allowed access to advanced U.S. aeronautical technology which made it possible to design and produce the Saab 37 Viggen much faster and cheaper than would otherwise have been possible.

According to the doctoral research of Nils Bruzelius at the Swedish National Defence College, the reason for this officially unexplained U.S. support was the need to protect U.S. Polaris submarines deployed just outside the Swedish west coast against the threat of Soviet anti-submarine aircraft.

However, Bruzelius' theory have been thoroughly debunked by Simon Moores and Jerker Widén.

The Viggen was powered by a single Volvo RM8 turbofan. This was essentially a licence-built variant of the Pratt & Whitney JT8D engine that powered commercial airliners of the 1960s, with an afterburner added for the Viggen. The airframe also incorporated a thrust-reverser to use during landings and land manoeuvres, which, combined with the aircraft having flight capabilities approaching a limited STOL-like performance, enabled operations from 500 m airstrips with minimal support.

The thrust reverser could be pre-selected in the air to engage when the nose-wheel strut was compressed after touchdown. The Viggen was the first aircraft to feature both afterburners and thrust-reversers. Only the Viggen, Concorde and the Panavia Tornado featured both afterburners and thrust-reversers.

The requirements from the Swedish Air Force dictated Mach 2 capability at high altitude and Mach 1 at low altitude.

At the same time, short-field takeoff and landing performance was also required.

Since the Viggen was developed initially as an attack aircraft instead of an interceptor (the Saab 35 Draken fulfilled this role), some emphasis was given to low fuel consumption at high subsonic speeds at low level for good range.

With turbofan engines just emerging and indicating better fuel economy for cruise than turbojet engines, the former was favoured, since the latter were mainly limited by metallurgy development resulting from limitations in turbine temperature.

Mechanical simplicity was also favoured, so the air intakes were simple D-section types with boundary layer splitter plates, while the fixed inlet had no adjustable geometry for improved pressure recovery.

The disadvantage was that the required engine would be very large. In fact, at the time of introduction, it was the second largest fighter engine, with a length of 6.1 m and 1.35 m diameter; only the Tumansky R-15 was bigger.

Saab had originally wanted the Pratt & Whitney TF30 as the Viggen's powerplant.

Since the engine design had not been completed in 1962 when the airframe vs. engine design size needed to be frozen, the JT8D was chosen as the basis for modification instead.

The RM8 became the second operational afterburning turbofan in the world, and also the first equipped with a thrust reverser. It had a bypass ratio of around 1.07:1 in the RM8A, which reduced to 0.97:1 in the RM8B.

The AJ, SF, SH and SK 37 models of the Viggen had the first version of the RM8A engine with uprated internal components from the JT8D that it was based on. Thrust was 65.6 kN dry and 115.6 kN with afterburner. For the JA 37, the RM8A was modified to an 8B by replacing one LP compressor stage with a fan stage and improved combustor, turbine and afterburner. Thrust is 72.1 kN dry and 125.0 kN with afterburner.

The engine was started via a small gas turbine, itself started by an electric motor. Standby power and cooling air for onboard avionics were supplied via an external cart. An internal battery permitted start of the starter turbine and main engine in absence of the standby power cart.

With the performance requirements to a large extent dictating the choice of the engine, the airframe turned out to be quite bulky compared to contemporary slimmer designs with turbojet engines. The first prototypes had a straight midsection fuselage that was later improved with a "hump" on the dorsal spine for reduced drag according to the area rule.

The wing had the shape of a double delta with a dogtooth added to improve longitudinal stability when carrying external stores. Each dogtooth was also used as a fairing for a radar warning receiver (RWR) antenna.

A consequence of a tailless delta design, such as in the Viggen, is that the elevons, which replace more conventional control surfaces, operate with a small effective moment arm; their use adds substantial weight to the aircraft at takeoff and landing. Hinged leading edge surfaces can help counteract this, but an even more effective tool is the canard. The canards were positioned behind the inlets and placed slightly higher than the main wing, but were not movable as control surfaces (however, they were equipped with flaps). The purpose of the canard wings were to act as vortex generators for the main wing and therefore provide more lift. An added benefit was that they also improved roll stability in the transonic region around Mach 0.9. The canard flaps were deployed in conjunction with the landing gear to provide even more lift for takeoff and landing.

To withstand the stresses of no-flare landings, Saab made extensive use of titanium in the construction of the Viggen, especially in the fuselage, and incorporated an unusual arrangement for the main landing gear, in which the two wheels on each leg were placed in tandem. While such a layout is common in airliners and cargo aircraft, it is rare in fighters, but allows stowage in a thinner wing.

The tall single vertical stabilizer (45 degrees sweepback on the leading edge) was foldable to make it easier to store in hangars. After prototype testing of the SK variant, reduced longitudinal stability was discovered. To correct this, the vertical stabilizer was extended 10 cm (4 in) and the pitot tube was moved from the top of the fin leading edge to about midpoint where a sawtooth was also incorporated. The JA model later used the same improvements.

The six tanks in the fuselage and wings held approximately 5,000 litres of fuel with an additional 1,500 litres in an external drop tank. The specific fuel consumption was only 0.63 for cruise speeds (fuel consumption was rated 18 mg/Ns dry and 71 with afterburner). The Viggen's consumption was around 15 kg/sec at maximum afterburner, which meant that the internal fuel was exhausted in just seven minutes due to the relative inefficiency of the turbofan over a turbojet at full afterburner. Performance comparisons with other aircraft from the same age are however slightly difficult, since no other fighter or attack aircraft aside from the Harrier and Yak-38 were designed for STOL or VTOL capability.

In the early 1960s, it was decided that the Viggen should be a single seat aircraft. A digital central computer and a head-up display replaced the human navigator. This computer, called CK 37 (centralkalkylator 37), was the world's first airborne computer to use integrated circuits. It utilized the STRIL 60 system to be linked with the Swedish defence systems. The main sensor was an Ericsson PS 37 X-band monopulse radar with several functions: air-to-ground and air-to-air telemetry and cartography. A Honeywell radar altimeter with transmitter and receiver in the canard wings was used to assist low altitude flight. A Decca Type 72 doppler navigation radar and a series of other electronic sub-systems were also provided. A novel landing-aid device, the TILS (Tactical Instrument Landing System), made by Cutler-Hammer AIL, was used to improve landing accuracy down to 30 m from the threshold on the short highway airbase system. ECM consisted of a Satt Elektronik radar warning receiver system in the wings and the tail, an optional Ericsson Erijammer pod and BOZ-100 chaff/flare pod. In total, the electronics weighed 600 kg which was a substantial amount for a single-engine, late 1960s fighter.

The SK 37 trainer omitted the radar and CK 37 navigational computer, navigating only using the Decca system and later DME. The radar warning receiver electronics were also removed.

Initially, only a single reconnaissance (S) variant was considered, but fitting cameras as well as a radar proved to be impossible. The SH 37 maritime strike and reconnaissance variant was very similar to the AJ 37 and differed mainly in a maritime optimized PS 371/A radar with longer range and cockpit camera and tape recorder for mission analysis. "Red Baron" and LOROP camera pods were usually carried on the fuselage pylons. The centerline fuel tank was converted for a short period of time to a camera pod with two Recon/Optical CA-200 1676 mm cameras. In addition to the reconnaissance equipment, the SH 37 could also use all weapons for the AJ 37.

For the photographic SF version, the radar in the nose was omitted in favour of one SKa 24 57 mm, three SKa 24C 120 mm and two SKa 31 600 mm photographic cameras as well as one VKa 702 Infrared linescan camera. The "Red Baron" and LOROP camera pods could also be carried on the fuselage pylons.

The avionics suite of the JA was a major improvement over the other variants designed a decade earlier. The onboard computer was a Singer-Kearfott SKC-2037 built under license by Saab as CD 107, a Garrett AiResearch LD-5 air data computer (also used in the F-14 Tomcat), a Saab-Honeywell SA07 automatic flight control systems (which was the first digital variant to enter production) and a KTL-70L inertial navigation system. In the cockpit, several dial-indicator instruments were replaced by two CRT displays; one target indicator MI (sw: MålIndikator) in the center and one tactical moving- and rotating map indicator TI (sw: Taktisk Indikator) to the right while the head-up display SI (sw: SiktlinjesIndikator - line-of-sight indicator) was retained.

The radar on the JA 37 was upgraded to a multi-mode, pulse-Doppler Ericsson PS 46/A unit more optimized for the fighter/interceptor role. It sported lookdown/shootdown capability, range up to 48 km (30 mi), continuous-wave illumination for the Skyflash missiles as well as the ability to track two targets while scanning. The MTBF was reported as 100 hours, a very high reliability level for that generation of avionics systems.

In 1992, an upgrade program of some of the AJ/SF/SH (with least hours on the airframe) to AJS/AJSF/AJSH was initiated because of delays of the new JAS 39 Gripen. The modifications were not too extensive and consisted to the major part of a new Ericsson computer processor system, MIL-STD-1553B databus and MIL-STD-1760 stores interface system to carry the Rb 15F anti-ship missile and DWS 39 Mjölner submunitions dispenser. An upgraded radar warning receiver system with recording capability as well as a Mission planning system via a portable cartridge were also implemented. The original PS 37/A radar from the AJ 37 was upgraded to the PS 371/A (from the SH 37) allowing the new AJS 37 to perform radar reconnaissance missions. No airframe- and very minor cockpit modifications were made.

The JA 37 was continuously upgraded throughout its lifetime. In 1985, the "fighter link" went into service, permitting encrypted data communication between four fighters and ground radar based fighter command. This enabled one fighter to "paint" an airborne enemy with guidance radar for the Skyflash missiles of the three other fighters in a group while they had their search and guidance radar switched off. This system was operational ten years before any other country's. The autopilot was also slaved to the radar control to obtain better precision firing the cannon.

In 1990, the PS 46/A was upgraded with higher resistance to jamming and the ability to track several targets at the same time. In 1993, the ability to generate virtual targets in the radar reduced the cost of flying aggressors for training.

Between 1992 and 1997, a major avionics upgrade program to the JA was implemented, given the new designation JA 37D. It consisted of an Ericsson CD207 mission computer, an ANP-37 stores management computer, linked via dual MIL-STD-1553B databuses permitting use of the Rb 99 AMRAAM. In the cockpit, a TI 327 color tactical moving-map display (originally intended for the Gripen) and a Synthetic Attitude Heading Reference System were installed. The ECM and ECCM suite were enhanced with improved electronics, upgraded radar warning receivers, a new Ericsson U95 jammer pod as well as the ability to carry BOY-401 chaff/flare dispensers on a separate location from the weapon pylons.

Between 1998 and 2000, the conversion of ten Sk 37 trainers to Sk 37E electronic aggressors was completed. The fairly substantial upgrade package consisted of the nose-radome mounted G24 jammer inherited from the decommissioned J 32E Lansen, U22/A jammer- and KB chaff/flare pods and radar warning receivers from the AJS 37 and a new U95 jammer pod all linked together with the MIL-STD-1553B databus. The rear cockpit for the Electronic Warfare Officer was improved with new displays and controls while retaining the ability to convert back to the original flight training role.

A weapons load of up to 7,000 kg could be accommodated on seven hardpoints; one centerline pylon, two fuselage pylons, two inner and two outer wing pylons. The centerline pylon was the only wet pylon and was usually occupied by an external fuel tank. The outboard wing pylons were never used in peacetime since aerodynamic flutter loads would structurally fatigue the wing.

The AJ 37 was designed to carry two RB 04E anti-ship missiles on the inboard wing pylons with an optional third missile on the centerline pylon.

An optional load consisted of two RB 05A air-to-surface missiles on the fuselage pylons. The RB 05A was later replaced by Rb 75 TV-guided missiles. In a ground-attack role, a combination of unguided 135 mm rockets in sextuple pods and 120 kg fragmentation bombs on quadruple-mounts could be used. Self-defense was provided with either ECM or 30 mm ADEN cannon pods with 150 rounds of ammunition on the inboard wing pylons.

Rockets had warheads of several types: the 50 mm M56GP 4 kg armour-piercing, the M56B with 6.9 kg of HE, and the M70 with a 4.7 kg HEAT warhead.

For the secondary air-to-air role and self defence, the Rb 28 IR-missile was initially selected, but was never used due to poor performance. This left the outboard wing pylons unutilised as the Rb 28 was the only missile integrated there up until the AJS modernisation. Instead, Rb 24/Rb 24J were used on the fuselage pylons and inboard wing pylons or in combination with optional 30 mm underwing ADEN cannon pods.

AJ 37 was under consideration as a carrier of both nuclear and chemical weapons, although no nuclear or chemical weapons were adopted by Sweden.

The SH 37 was capable of carrying the same configuration of weapons as the AJ 37. However, since it was only used in the maritime role, only the RB 04E in combination with Rb 24/Rb 24J for self defense were employed. The chaff and jammer-pods was the most commonly used load.

Both the SF and SK variants lacked the radar and could not carry the guided air-to-surface missiles as the AJ and SH. The SF could carry Rb 24/Rb 24J for self defense though. The unguided cannon and rocket pods were also an option.

With the introduction of the JA 37 in 1979 came the Ericsson PS 46/A radar capable of guiding the two semi-active radar homing Rb 71 missiles on the fuselage pylons simultaneously in combination with Rb 24J/Rb 24J air-to-air missiles. Unlike the strike variant a KCA 30mm Oerlikon internal cannon was carried as well as 126 rounds, in a conformal pod under the fuselage. The firing rate was selectable at 22 or 11 rounds. The KCA cannon fired 50% heavier shells at higher velocity than the older ADENs, giving a much higher kinetic energy. This, in conjunction with the fire control system, allowed air-to-air engagements at longer range than other fighters. The unguided cannon and rocket pods were available in the secondary ground-attack role.

The centerline pylon was almost exclusively carried a semi-permanent fuel tank, which was jettisonable in the event of a dogfight.

In 1987, the more advanced all-aspect Rb 74 air-to-air missile was introduced for the JA 37. With the major upgrade of the JA to JA 37D in 1997 came the ability to carry four Rb 99 on the fuselage- and inner wing pylons. In addition, a U95 ECM pod could now be carried under the right wing in place of an AMRAAM as well as chaff and flare dispensers on a pair of hitherto unused pylons just behind the main landing gear on each wing.

With the extensive electronics upgrade of the old AJ/SF/SH in 1992 came the ability to carry the new Rb 74 on all weapons pylons. The AJS and AJSH also received the Rb 15F anti-ship missile and BK 90 stand-off cluster bomb originally intended for the delayed JAS 39 Gripen.

One hundred and ten of the original, ground-attack optimized variant, AJ 37 were built, with the first operational squadron established in 1972 at F 7 Såtenäs.

A two-seat trainer was not initially planned since it was considered that new pilots could get enough experience with delta-winged aircraft on the SK 35 Draken trainer. Eventually, however, 18 SK 37 two-seat trainers were ordered and delivered in 1973. To make room for the second cockpit, one fuel tank and some avionics were removed. The radar was also omitted limiting the weapons load to gun pods and unguided rockets.

A total of 26 of the SH 37 maritime reconnaissance and strike variant were built in 1974, replacing the S 32C Lansen. Although fitted with radar and weaponry, the SH 37 Viggen could also undertake photographic missions with its single long-range camera, while external pods could carry a photographic day-set, a "Red Baron" IR set, an ELINT set, and AQ series ECM (made by SATT).

A further 26 of the SF 37 reconnaissance variant were also delivered to replace the S 35 Draken in 1975. These were recognizable by having an elongated nose, equipped with six cameras and a VKa 702 infrared linescanner for night reconnaissance. Also, the "Red Baron" pod, with three IR cameras was widely used, as well as an ELINT set.

Although the Viggen was offered for sale worldwide, and regarded as a very competent aircraft, no export sales occurred. Reasons to explain Saab's failure to sell a competitively priced, highly advanced and well-respected aircraft include the Swedish government's relatively strict controls on arms exports to undemocratic countries, potential customers' doubts about continuity of support and supply of spare parts in the event of a conflict disapproved of by Sweden, and strong diplomatic pressure of larger nations. The United States blocked an export of Viggens to India in 1978 by not issuing an export license for the RM8/JT8D engine, forcing India to choose the SEPECAT Jaguar instead.

The Viggen saw initial service in natural metal, later receiving an extremely elaborate disruptive camouflage scheme for the AJ/SF/SH/SK variants and the first 27 JA aircraft. The 28th JA was painted in a gray tone that turned out too close to white. All latter JA aircraft were painted in a darker light/dark gray, appropriate for a high altitude fighter.

The final Viggen production variant was the JA 37 interceptor entering service in 1980.

The last of 149 JA 37s was delivered in 1990. Differences from the previous models included an improved and more powerful RM8B engine, a new PS 46/A interception radar, new computers, HUD, ECM and some other subsystems.

Unusually for a 1970s fighter, three multi-purpose CRT display screens were fitted within the cockpit, in a system called AP-12, that also included a new model of HUD.

The new radar was compatible with the Skyflash medium-range missiles, for the first time in a Swedish fighter.

Two Skyflash missiles could be carried under the wings on hardpoints, as well as four Sidewinder J or L models.

Another improvement was the addition of an Oerlikon KCA 30 mm cannon mounted internally, with 126 rounds of 360 g ammunition.

The structural strength was also improved, especially for the multi-sparred wings (initially Viggens had a high loss rate, with 21 aircraft lost in the early years).

Various upgrades have been performed over the years, mainly to cockpit equipment, weapons and sensor fit. Between 1998 and 2000, ten SK 37 trainers were converted to SK 37E electronic warfare trainers to replace the aging J 32E Lansen.

The SEPECAT Jaguar (fig. 50) is an Anglo-French jet ground attack aircraft, originally used by the British Royal Air Force and the French Armée de l'Air in the close air support and nuclear strike role, and still in service with several export customers, notably the Indian Air Force and the Royal Air Force of Oman.

Originally conceived in the 1960s as jet trainer with a light ground attack capability, the requirement for the aircraft soon changed to include supersonic performance, reconnaissance and tactical nuclear strike roles.

A carrier-based variant was also planned for French service, but this was cancelled in favour of the cheaper Dassault Super Étendard.

The airframes were manufactured by SEPECAT, a joint venture between Breguet and the British Aircraft Corporation, one of the first major joint-Anglo-French military aircraft programs.

The Jaguar was successfully exported to India, Oman, Ecuador and Nigeria. With various airforces, the Jaguar was used in numerous conflicts and military operations in Mauritania, Chad, Iraq, Bosnia, and Pakistan, as well as providing a ready nuclear delivery platform for Britain, France, and India throughout the latter half of the Cold War and beyond. In the Gulf War, the Jaguar was praised for its reliability and was a valuable coalition resource. The aircraft served with the Armée de l'Air as the main strike/attack aircraft until 1 July 2005, and with the Royal Air Force until the end of April 2007. It was replaced by the Panavia Tornado and the Eurofighter Typhoon in the RAF and the Dassault Rafale in the Armée de l'Air. India plans in the long term to replace its Jaguar fleet with the developing Advanced Medium Combat Aircraft (AMCA).

***Fig. 50.** The SEPECAT Jaguar.*

The T-45 Goshawk (fig. 51) is a highly modified version of the BAE Hawk land-based training jet aircraft. Manufactured by

McDonnell Douglas (now Boeing) and British Aerospace (now BAE Systems), the T-45 is used by the United States Navy as an aircraft carrier-capable trainer.

Fig. 51. *The T-45 Goshawk.*

The T-45 Goshawk is a fully carrier-capable version of the Hawk Mk.60.

It was developed for the United States Navy (USN) and United States Marine Corps (USMC) jet flight training.

The Goshawk's origins began in the mid-1970s, when the US Navy began looking for replacement for its T-2 and TA-4 trainers. The US Navy started the VTXTS advanced trainer program in 1978. British Aerospace and McDonnell Douglas proposed a version of the Hawk and were awarded the T-45 contract in 1981.

The Hawk had not been designed for carrier operations and numerous modifications were required for Navy carrier use. These included improvements to the low-speed handling characteristics and a reduction in the approach speed. Other changes were strengthened airframe, more robust and wider landing gear with catapult tow bar attachment and an arresting hook. It features a two-wheel nose landing gear.

The Goshawk first flew in 1988 and became operational in 1991. BAE Systems manufactures the fuselage aft of the cockpit, the air inlets, the vertical stabilizer of the T-45 at Samlesbury, and the wings at Brough, England. Boeing, which merged with McDonnell Douglas in 1997, manufactures the remainder of the aircraft and assembles them in St. Louis, Missouri.

On 16 March 2007 the 200th airframe was delivered to the US Navy. Their requirements call for 223 aircraft, and the T-45 is slated to continue in service until at least 2035.

The T-45 has been used for intermediate and advanced portions of the Navy/Marine Corps Student Naval Aviator strike pilot training program with Training Air Wing One at Naval Air Station Meridian, Mississippi and Training Air Wing Two at Naval Air Station Kingsville, Texas. The T-45 replaced the T-2C Buckeye intermediate jet trainer and the TA-4J Skyhawk II advanced jet trainer with an integrated training system that includes the T-45 Goshawk aircraft, operational and instrument flight simulators (OFT/IFT), academics, and training integration system support.

In 2008, the T-45C also began operation in the advanced portion of Navy/Marine Corps Student Naval Flight Officer (NFO) training with Training Air Wing Six at Naval Air Station Pensacola, Florida.

The T-45's A and C models are currently in operational use. The T-45A, which became operational in 1991, contains an analog cockpit design while the newer T-45C, which was first delivered in December 1997, features a new digital "glass cockpit" design. All T-45A aircraft will eventually be converted to a T-45C configuration under the T-45 Required Avionics Modernization Program (T-45 RAMP).

The Panavia Tornado (fig. 52) is a family of twin-engine, variable-sweep wing combat aircraft, which was jointly developed by the United Kingdom, West Germany and Italy. There are three primary versions of the Tornado; the Tornado IDS (interdictor/strike) fighter-bomber, the suppression of enemy air defences Tornado ECR (electronic combat/reconnaissance) and the Tornado ADV (air defence variant) interceptor.

Developed and built by Panavia, a tri-national consortium consisting of British Aerospace (previously British Aircraft

Corporation), MBB of West Germany, and Aeritalia of Italy, the Tornado first flew on 14 August 1974, and saw action with the Royal Air Force (RAF), Italian Air Force and Royal Saudi Air Force in the Gulf War. International co-operation continued after its entry into service within the Tri-National Tornado Training Establishment, a tri-nation training and evaluation unit operating from RAF Cottesmore, UK. Including all variants, 992 aircraft were built for the three partner nations and Saudi Arabia.

Fig. 52. The Panavia Tornado.

The Northrop/McDonnell Douglas YF-23 (fig. 53) was a prototype fighter aircraft designed for the United States Air Force. The YF-23 was a finalist in the U.S. Air Force's Advanced Tactical Fighter competition. Two YF-23s were built and were nicknamed "Black Widow II" and "Gray Ghost". The YF-23 lost the contest to the Lockheed YF-22, which entered production as the Lockheed Martin F-22 Raptor.

Fig. 53. *The Northrop/McDonnell Douglas YF-23 Black Widow II.*

The YF-22 and YF-23 were competing in the USAF's Advanced Tactical Fighter (ATF) program, conceived in the early 1980s, to provide a replacement for the F-15 Eagle. Contracts for the two most promising designs were awarded in 1986.

The YF-23 was designed to meet USAF requirements for survivability, supersonic cruise (supercruise), stealth, and ease of maintenance. Designed with all-aspect stealth as a high priority, Northrop drew on the company's experience with the B-2 Spirit and F/A-18 Hornet. The YF-23 was an unconventional-looking aircraft with trapezoidal wings, substantial area-ruling, and a V-tail. Similar to the B-2, the exhaust from the YF-23's engines flows through troughs lined with heat ablating tiles, which shields the exhaust from infrared (IR) missile detection from below. The vehicle management system coordinates movements of the control surfaces for maneuvers and for stable flight, along with other aircraft functions. The wing flaps and ailerons deflect inversely on either side to provide roll. Pitch was provided by movement of both V-tails, and yaw was supplied by opposite movement. Deflecting the wing flaps down and ailerons up on both sides simultaneously provided for aerodynamic braking.

Although possessing an advanced design, in order to reduce costs and development, a number of F-15 Eagle components were utilized including the standard F-15 nose wheel unit and the forward cockpit of the F-15E Strike Eagle. Two aircraft were built. YF-23 I (PAV-1) was fitted with Pratt & Whitney YF119 engines, while YF-23 II (PAV-2) was fitted with General Electric YF120 engines. The YF-23 featured fixed nozzles. The first YF-23 was rolled out on 22 June 1990, and first flew on 27 August 1990. YF-23 II first flew on 26 October 1990.

The black YF-23 (PAV-1) was nicknamed "Black Widow II", after the Northrop P-61 Black Widow of World War II and had a red hourglass marking resembling the underbelly marking of the black widow spider. The black widow marking was briefly seen under PAV-1 before being removed at the insistence of Northrop management. The gray colored YF-23 (PAV-2) was nicknamed "Gray Ghost".

Both YF-23s were furnished in the configuration specified before the requirement for thrust reversing was dropped. The weapons bay was configured for weapons launch but no missiles were fired, unlike Lockheed's demonstration aircraft. The YF-23s flew 50 times for a total of 65.2 hours. The first YF-23 with P&W engines supercruised at Mach 1.43 on 18 September 1990 and the second YF-23 with GE engines reached Mach 1.6 on 29 November 1990. For comparison, the YF-22 achieved Mach 1.58 in supercruise. The flight testing demonstrated Northrop's predicted performance values for the YF-23.

The YF-22 won the competition in April 1991. The YF-23 design was more stealthy and faster, but the YF-22 was more agile. It has been speculated in the aviation press that the YF-22 was also seen as more adaptable to the Navy's Navalized Advanced Tactical Fighter (NATF), though as it turned out the US Navy abandoned NATF a few months later.

After losing the competition, both YF-23s were transferred to NASA's Dryden Flight Research Center, at Edwards AFB, California without the engines. NASA planned to use one of the aircraft to study strain gauge loads calibration techniques, but this did not occur.

In late 2004, Northrop Grumman proposed a YF-23 based design for the USAF's interim bomber requirement, a role for which the FB-22 and B-1R are also competing. Aircraft PAV-2 was modified by Northrop as a full size model of its proposed interim bomber. The interim bomber requirement has since been canceled in favor of a more long-term, bomber replacement requirement. The same YF-23-derived design could possibly be adapted to fulfill this role as well. However, it appears the possibility of a YF-23-based interim bomber was ended with the 2006 Quadrennial Defense Review, in favor of a long range bomber with a much greater range.

Most helicopters have a single main rotor, but torque created as the engine turns the rotor against its air drag causes the body of the helicopter to turn in the opposite direction to the rotor. To eliminate this effect, some sort of antitorque control must be used. The design that Igor Sikorsky settled on for his VS-300 was a smaller rotor mounted vertically on the tail. The tail rotor pushes or pulls against the tail to counter the torque effect, and has become the recognized convention for helicopter design. Some helicopters utilize alternate antitorque controls in place of the tail rotor, such as the ducted fan (called Fenestron or FANTAIL), and NOTAR. NOTAR provides antitorque similar to the way a wing develops lift, through the use of a Coandă effect on the tailboom (fig. 54).

Fig. 54. *MD Helicopters 520N NOTAR.*

The use of two or more horizontal rotors turning in opposite directions is another configuration used to counteract the effects of torque on the aircraft without relying on an antitorque tail rotor. This allows the power normally required to drive the tail rotor to be applied to the main rotors, increasing the aircraft's lifting capacity. Primarily, there are three common configurations that use the counterrotating effect to benefit the rotorcraft. Tandem rotors are two rotors with one mounted behind the other. Coaxial rotors are two rotors that are mounted one above the other with the same axis. Intermeshing rotors are two rotors that are mounted close to each other at a sufficient angle to allow the rotors to intermesh over the top of the aircraft. Transverse rotors is another configuration found on tiltrotors and some earlier helicopters, where the pair of rotors are mounted at each end of the wings or outrigger structures. Tip jet designs permit the rotor to push itself through the air, and avoid generating torque (fig. 55).

Fig. 55. The Piasecki H 21 employed tandem rotors.

The number, size and type of engine used on a helicopter determines the size, function and capability of that helicopter design. The earliest helicopter engines were simple mechanical devices, such as rubber bands or spindles, which relegated the size of helicopters to toys and small models. For a half century before the first airplane flight, steam engines were used to forward the development of the understanding of helicopter aerodynamics, but the limited power did not allow for manned flight.

The introduction of the internal combustion engine at the end of the 19th century became the watershed for helicopter development as engines began to be developed and produced that were powerful enough to allow for helicopters able to lift humans.

Early helicopter designs utilized custom-built engines or rotary engines designed for airplanes, but these were soon replaced by more powerful automobile engines and radial engines.

The single, most-limiting factor of helicopter development during the first half of the 20th century was that the amount of power produced by an engine was not able to overcome the engine's weight in vertical flight. This was overcome in early successful helicopters by using the smallest engines available. When the compact, flat engine was developed, the helicopter industry found a lighter-weight powerplant easily adapted to small helicopters, although radial engines continued to be used for larger helicopters.

Turbine engines revolutionized the aviation industry, and the turboshaft engine finally gave helicopters an engine with a large amount of power and a low weight penalty. The turboshaft engine was able to be scaled to the size of the helicopter being designed, so that all but the lightest of helicopter models are powered by turbine engines today.

Special jet engines developed to drive the rotor from the rotor tips are referred to as tip jets. Tip jets powered by a remote compressor are referred to as cold tip jets, while those powered by combustion exhaust are referred to as hot tip jets. An example of a cold jet helicopter is the Sud-Ouest Djinn, and an example of the hot tip jet helicopter is the YH-32 Hornet.

Some radio-controlled helicopters and smaller, helicopter-type unmanned aerial vehicles, use electric motors. Radio-controlled helicopters may also have piston engines that use fuels other than gasoline, such as Nitromethane. Some turbine engines commonly used in helicopters can also use biodiesel instead of jet fuel.

A helicopter has four flight control inputs. These are the cyclic, the collective, the anti-torque pedals, and the throttle. The cyclic control is usually located between the pilot's legs and is commonly called the cyclic stick or just cyclic. On most helicopters, the cyclic is similar to a joystick. However, the Robinson R22 and Robinson R44 have a unique teetering bar cyclic control system and a few helicopters have a cyclic control that descends into the cockpit from overhead.

The control is called the cyclic because it changes the pitch of the rotor blades cyclically. The result is to tilt the rotor disk in a particular direction, resulting in the helicopter moving in that direction. If the pilot pushes the cyclic forward, the rotor disk tilts forward, and the rotor produces a thrust in the forward direction. If the pilot pushes the cyclic to the side, the rotor disk tilts to that side and produces thrust in that direction, causing the helicopter to hover sideways.

The collective pitch control or collective is located on the left side of the pilot's seat with a settable friction control to prevent inadvertent movement. The collective changes the pitch angle of all the main rotor blades collectively (i.e. all at the same time) and independently of their position. Therefore, if a collective input is made, all the blades change equally, and the result is the helicopter increasing or decreasing in altitude.

The anti-torque pedals are located in the same position as the rudder pedals in a fixed-wing aircraft, and serve a similar purpose, namely to control the direction in which the nose of the aircraft is pointed. Application of the pedal in a given direction changes the pitch of the tail rotor blades, increasing or reducing the thrust produced by the tail rotor and causing the nose to yaw in the direction of the applied pedal. The pedals mechanically change the pitch of the tail rotor altering the amount of thrust produced.

Helicopter rotors are designed to operate in a narrow range of RPM. The throttle controls the power produced by the engine, which is connected to the rotor by a fixed ratio transmission. The purpose of the throttle is to maintain enough engine power to keep the rotor RPM within allowable limits in order to keep the rotor producing enough lift for flight. In single-engine helicopters, the throttle control is a motorcycle-style twist grip mounted on the collective control, while dual-engine helicopters have a power lever for each engine.

A Swashplate transmits the pilot commands to the main rotor blades for articulated rotors.

There are three basic flight conditions for a helicopter; hover and forward flight, and the transition between the two..

Hovering is the most challenging part of flying a helicopter. This is because a helicopter generates its own gusty air while in a hover, which acts against the fuselage and flight control surfaces. The end result is constant control inputs and corrections by the pilot to keep the helicopter where it is required to be. Despite the complexity of the task, the control inputs in a hover are simple. The cyclic is used to eliminate drift in the horizontal plane, that is to control forward and back, right and left. The collective is used to maintain altitude. The pedals are used to control nose direction or heading. It is the interaction of these controls that makes hovering so difficult, since an adjustment in any one control requires an adjustment of the other two, creating a cycle of constant correction.

Transition from hover to forward flights

As a helicopter moves from hover to forward flight it enters a state called Translational lift which provides extra lift without increasing power. This state, most typically, occurs when the airspeed reaches approximately 16-24 knots, and may be necessary for a helicopter to obtain flight.

In forward flight a helicopter's flight controls behave more like that in a fixed-wing aircraft. Displacing the cyclic forward will cause the nose to pitch down, with a resultant increase in airspeed and loss of altitude. Aft cyclic will cause the nose to pitch up, slowing the helicopter and causing it to climb. Increasing collective (power) while maintaining a constant airspeed will induce a climb while decreasing collective will cause a descent. Coordinating these two inputs, down collective plus aft cyclic or up collective plus forward cyclic, will result in airspeed changes while maintaining a constant altitude. The pedals serve the same function in both a helicopter and a fixed-wing aircraft, to maintain balanced flight. This is done by applying a pedal input in whichever direction is necessary to center the ball in the turn and bank indicator.

The main limitation of the helicopter is its low speed. There are several reasons a helicopter cannot fly as fast as a fixed wing aircraft. When the helicopter is hovering, the outer tips of the rotor travel at a speed determined by the length of the blade and the RPM. In a moving helicopter, however, the speed of the blades relative to the air depends on the speed of the helicopter as well as on their rotational velocity. The airspeed of the advancing rotor blade is much higher than that of the helicopter itself. It is possible for this blade to exceed the speed of sound, and thus produce vastly increased drag and vibration. See Wave drag.

Because the advancing blade has higher airspeed than the retreating blade and generates a dissymmetry of lift, rotor blades are designed to "flap" – lift and twist in such a way that the advancing blade flaps up and develops a smaller angle of attack. Conversely, the retreating blade flaps down, develops a higher angle of attack, and generates more lift. At high speeds, the force on the rotors is such that they "flap" excessively and the retreating blade can reach too high an angle and stall. For this reason, the maximum safe forward airspeed of a helicopter is given a design rating called VNE, Velocity, Never Exceed. In addition it is possible for the helicopter to fly at an airspeed where an excessive amount of the retreating blade stalls which results in high vibration, pitch -up, and roll into the retreating blade.

During the closing years of the 20th century designers began working on helicopter noise reduction. Urban communities have often expressed great dislike of noisy aircraft, and police and passenger helicopters can be unpopular. The redesigns followed the closure of some city heliports and government action to constrain flight paths in national parks and other places of natural beauty.

Helicopters also vibrate; an unadjusted helicopter can easily vibrate so much that it will shake itself apart. To reduce vibration, all helicopters have rotor adjustments for height and weight. Blade height is adjusted by changing the pitch of the blade. Weight is adjusted by adding or removing weights on the rotor head and/or at the blade end caps. Most also have vibration dampers for height and pitch. Some also use mechanical feedback systems to sense and counter vibration. Usually the feedback system uses a mass as a "stable reference" and a linkage from the mass operates a flap to adjust the rotor's angle of attack to counter the vibration. Adjustment is difficult in part because measurement of the vibration is hard, usually requiring sophisticated accelerometers mounted throughout the airframe and gearboxes. The most common blade vibration adjustment measurement system is to use a stroboscopic flash lamp, and observe painted markings or coloured reflectors on the underside of the rotor blades. The traditional low-tech system is to mount coloured chalk on the rotor tips, and see how they mark a linen sheet. Gearbox vibration most often requires a gearbox overhaul or replacement. Gearbox or drive train vibrations can be extremely harmful to a pilot. The most severe being pain, numbness, loss of tactile discrimination and dexterity.

Canada Receives First CH-148 Maritime Helicopter.

The first interim maritime helicopter, the CH-148 Cyclone, arrived at 12 Wing Shearwater, Nova Scotia, last week to support training of Canadian Forces (CF) aircrew and technicians for the Maritime Helicopter Project (fig. 56).

"The arrival of this helicopter in Shearwater demonstrates progress with this project and brings us one step closer towards the delivery of a Maritime Helicopter capability that provides the Canadian Forces with a modern, flexible helicopter to assist in the defence of Canada and Canadian interests well into the future," said the Honourable Peter MacKay, Minister of National Defence.

On May 13, 2011, Sikorsky Operations International Inc. flew the helicopter to Shearwater as part of its contractual obligation under the Maritime Helicopter Project to deliver initial cadre training to the Canadian Forces. The helicopter will be used as a ground-based training aid for technicians.

The arrival of this helicopter does not mark formal delivery at this time as Sikorsky has not yet met all of the contractual delivery requirements. The helicopter will remain under Sikorsky title and control until all requirements are met.

The CF will take formal delivery and assume ownership of the helicopter once a Canadian military airworthiness certificate is granted and once initial aircrew flight training is conducted.

Sikorsky continues to make steady progress. Formal delivery of the first interim maritime helicopter is expected later this summer. The new Cyclone, in its final configuration, will be at the forefront of modern technology and will be one of the most capable maritime helicopters in the world.

Fig. 56. CH-148 Cyclone.

Fig. 57. *Agusta A129 Mangusta.*

The Agusta A129 Mangusta (English: Mongoose) is an attack helicopter originally designed and produced by Agusta in Italy (fig. 57). It was the first attack helicopter to be designed and produced wholly in Western Europe. The TAI/AgustaWestland T-129 ATAK is an enhanced version of the A129, and its development is now the responsibility of Turkish Aerospace Industries (TAI), with AgustaWestland as the primary partner.

The Bell AH-1 SuperCobra is a twin-engine attack helicopter based on the US Army's AH-1 Cobra. The twin Cobra family includes the AH-1J SeaCobra, the AH-1T Improved SeaCobra, and the AH-1W SuperCobra. The AH-1W is the backbone of the United States Marine Corps's attack helicopter fleet, but will be replaced in service by the Bell AH-1Z Viper upgrade in the next decade (fig. 58).

Fig. 58. *The Bell AH-1W Super Cobra is the backbone of the United States Marine Corps's attack helicopter fleet.*

The AH-1 Cobra was developed in the mid-1960s as an interim gunship for the U.S. Army for use in Vietnam. The Cobra shared the proven transmission, rotor system, and the T53 turboshaft engine of the UH-1 "Huey". By June 1967, the first AH-1G HueyCobras had been delivered. Bell built 1,116 AH-1Gs for the U.S. Army between 1967 and 1973, and the Cobras chalked up over a million operational hours in Vietnam.

The U.S. Marine Corps was very interested in the AH-1G Cobra, but preferred a twin-engined version for improved safety in over-water operations, and also wanted a more potent turret-mounted weapon. At first, the Department of Defense had balked at providing the Marines with a twin-engined version of the Cobra, in the belief that commonality with Army AH-1Gs outweighed the advantages of a different engine fit. However, the Marines won out and awarded Bell a contract for 49 twin-engined AH-1J SeaCobras in May 1968. As an interim measure, the U.S. Army passed on 38 AH-1Gs to the Marines in 1969. The AH-1J also received a more powerful gun turret. It featured a three barrel 20 mm XM197 cannon that was based on the six barrel M61 Vulcan cannon.

The Marine Corps requested greater load carrying capability in high temperatures for the Cobra in the 1970s. Bell used systems from its Model 309 to develop the AH-1T. This version had a lengthened tailboom and fuselage with an upgraded transmission and engines from the 309. Bell designed the AH-1T to be more reliable and easier to maintain in the field. The version was given full TOW missile capability with targeting system and other sensors. An advanced version, known as the AH-1T+ with more powerful T700-GE-700 engines and advanced avionics was proposed to Iran in the late 1970s, but the overthrow of the Shah of Iran resulted in the sale being canceled.

In the early 1980s, the U.S. Marine Corps sought a new navalized helicopter, but was denied funding to buy the AH-64 Apache by Congress in 1981. The Marines in turn pursued a more powerful version of the AH-1T. Other changes included modified fire control systems to carry and fire AIM-9 Sidewinder and AGM-114 Hellfire missiles. The new version was funded by Congress and received the AH-1W designation. Deliveries of AH-1W SuperCobras totaled 179 new-built helicopters plus 43 upgrades of AH-1Ts.

The AH-1T+ demonstrator and AH-1W prototype was later tested with a new experimental composite four blade main rotor system. The new system offered better performance, reduced noise and improved battle damage tolerance. Lacking a USMC contract, Bell developed this new design into the AH-1Z with its own funds. By 1996, the Marines were again not allowed to order the AH-64. Developing a marine version of the Apache would have been expensive and it was likely that the Marine Corps would be its only customer. They instead signed a contract for upgrading 180 AH-1Ws into AH-1Zs.

The AH-1Z Viper features several design changes. The AH-1Z's two redesigned wing stubs are longer with each adding a wing-tip station for a missile such as the AIM-9 Sidewinder. Each wing has two other stations for 70 mm (2.75 in) Hydra rocket pods, or AGM-114 Hellfire quad missile launcher. The Longbow radar can be mounted on a wing tip station.

During the closing months of the United States' involvement in the Vietnam War, the Marine Corps embarked the AH-1J SeaCobra assigned to HMA-369 (now HMLA-369) aboard the USS Denver (LPD-9), USS Cleveland (LPD-7), and later the USS Dubuque (LPD-8), for sea-based interdiction of the Ho Chi Minh Trail in North Vietnam in the vicinity of Hon La (Tiger) Island. These were termed Marine Hunter-Killer (MARHUK) Operations and lasted from June to December 1972.

Marine Cobras took part in the invasion of Grenada, during Operation Urgent Fury in 1983, flying close-support and helicopter escort missions. Two Marine AH-1Ts were shot down and three crew members killed. The Marines also deployed the AH-1 off the coast of Beirut, Lebanon in 1983, during that nation's civil war. The AH-1s were armed with Sidewinder missiles and guns as an emergency air defense measure against the threat of light civil aircraft employed by suicide bombers.

USMC Cobras provided escort in the Persian Gulf in the late 1980s while the Iran–Iraq War was ongoing. The Cobras sank three Iranian patrol boats while losing one AH-1T to Iranian anti-aircraft fire. USMC Cobras from the USS Saipan (LHA-2) flew "top cover" during an evacuation of American and other foreign nationals from Liberia in 1990.

During the Gulf War, 78 Marine SuperCobras deployed, and flew a total of 1,273 sorties in Iraq with no combat losses. However, three AH-1s were lost to accidents during and after the combat operations. The AH-1W units were credited with destroying 97 tanks, 104 armored personal carriers and vehicles, and two anti-aircraft artillery sites during the 100-hour ground campaign.

Marine Cobras provided support for the US humanitarian intervention in Somalia, during Operation Restore Hope in 1992-1993. They were also employed during the U.S. invasion of Haiti in 1994. USMC Cobras were used in U.S. military interventions in the former Yugoslavia in the 1990s, and assisted in the rescue of USAF Captain Scott O'Grady, after his F-16 was shot down by a SAM in June 1995.

AH-1 Cobras continue to operate with the U.S. Marine Corps. USMC Cobras were also used in operations throughout the 1990s. USMC Cobras have also served in Operation Enduring Freedom in Afghanistan and in Operation Iraqi Freedom in the ongoing conflict in Iraq. While new replacement aircraft were considered as an alternative to major upgrades of the AH-1 fleet, Marine Corps studies showed that an upgrade was the most affordable, most supportable and most effective solution for the Marine Corps light attack helicopter mission.

In 1971, Iran purchased 202 improved AH-1J Cobras, with the name "AH-IJ International", from the United States. This improved Cobra, known as the AH-1J International, resulted from this contract featured an uprated P&WC T400-WV-402 engine and stronger drivetrain. Recoil damping gear was fitted to the 20mm gun turret, and the gunner was given a stabilized sight and even a stabilized chair. 62 of the International AH-1Js delivered to the Shah's forces were TOW-capable, while the rest were not.

They participated in the Iran–Iraq War. Iranian AH-1J SeaCobras engaged in air combat with Iraqi Mi-24s on several separate occasions during the war. The results of these engagements are disputed. One document cited that "Iranian AH-1Js engaged Iraqi MI-8 Hip and MI-24 Hind helicopters. Unclassified sources report that the Iranian AH-1 pilots achieved a 10:1 kill ratio over the Iraqi helicopter pilots during these engagements (1:5). Additionally, Iranian AH-1 and Iraqi fixed-wing aircraft engagements also occurred. Others claim that in the entire eight-year conflict, ten Iranian AH-1Js were lost in combat, compared to six Iraqi Mi-24s. The skirmishes are described as fairly evenly matched in another source. Iranian AH-1Js are still operating today and have undergone indigenous upgrade programs. In 1988, two Soviet MiG-23s shot down a pair of Iranian AH-1Js that had strayed into western Afghan airspace.

Turkey bought ten AH-1W SuperCobras in the early 1990s, and supplemented this fleet with 32 ex-US Army AH-1 Cobras.

The US Army Cobras included some TAH-1P trainers, while the rest were brought up to AH-1F standard. AH-1Ws have been used in the war against the Kurdistan Workers' Party, or PKK. Turkey seeks to purchase nine additional AH-1Ws, but the US seems unwilling to sell them.

However, Turkey is to obtain two AH-1Ws from the USMC inventory in 2011.

The Bell AH-1Z Viper (fig. 59) is a twin-engine attack helicopter based on the AH-1W SuperCobra, that was developed for the United States Marine Corps. The AH-1Z features a four-blade, bearingless, composite main rotor system, uprated transmission, and a new target sighting system. The AH-1Z is part of the H-1 upgrade program. It is also called "Zulu Cobra" in reference to its variant letter.

Fig. 59. *The Bell AH-1Z Viper.*

The MH-6 Little Bird (also known as Killer Egg) and its attack variant, the AH-6 (fig. 60), are single-engine light helicopters used for special operations aviation in the United States Army.

Originally based on a modified OH-6A, it was later based on the MD 500E, with a single five-bladed main rotor. The newest version, the MH-6M, is based on the MD 530F and has a single, six-bladed main rotor and four-bladed tail rotor.

Fig. 60. *The AH-6 Little Bird.*

The Boeing AH-64 Apache (fig. 61) is a four-blade, twin-engine attack helicopter with a tailwheel-type landing gear arrangement, and a tandem cockpit for a crew of two.

The Apache was developed as Model 77 by Hughes Helicopters for the United States Army's Advanced Attack Helicopter program to replace the AH-1 Cobra.

First flown on 30 September 1975, the AH-64 features a nose-mounted sensor suite for target acquisition and night vision systems.

The Apache is armed with a 30-millimeter (1.2 in) M230 Chain Gun carried between the main landing gear, under the aircraft's forward fuselage.

It has four hardpoints mounted on stub-wing pylons, typically carrying a mixture of AGM-114 Hellfire and Hydra 70 rocket pods. The AH-64 also features double- and triple-redundant aircraft systems to improve survivability for the aircraft and crew in combat, as well as improved crash survivability for the pilots.

Fig. 61. *The AH-64 Apache.*

The U.S. Army selected the AH-64 over the Bell YAH-63 in 1976, awarding Hughes Helicopters a pre-production contract for two more aircraft. In 1982, the Army approved full production. McDonnell Douglas continued production and development after purchasing Hughes Helicopters from Summa Corporation in 1984. The first production AH-64D Apache Longbow, a greatly upgraded version of the original Apache, was delivered to the Army in March 1997. AH-64 production is continued by the Boeing Defense, Space & Security division; over one thousand AH-64s have been produced to date.

The U.S. Army is the primary operator of the AH-64, however it has also become the primary attack helicopter of several nations it has been exported to, including Greece, Japan, Israel, the Netherlands and Singapore; as well as being produced under license in the United Kingdom as the AgustaWestland Apache. U.S. AH-64s have served in conflicts in Panama, Persian Gulf War, Kosovo War, Afghanistan, and Iraq. Israel has made active use of the Apache in its military conflicts in Lebanon and Gaza Strip; while two coalition allies have deployed their AH-64s in Afghanistan and Iraq.

Following the cancellation of the AH-56 Cheyenne in 1972, in favor of United States Air Force and Marine Corps projects like the A-10 Thunderbolt II and Harrier, the United States Army sought an aircraft to fill an anti-armor attack role that would still be under Army command; the 1948 Key West Agreement forbade the Army from owning fixed-wing aircraft. The Army wanted an aircraft better than the AH-1 Cobra in firepower, performance and range. It would have the maneuverability for terrain following nap-of-the-earth (NoE) flying. To this end, the US Army issued a Request For Proposals (RFP) for an Advanced Attack Helicopter (AAH) on 15 November 1972. As a sign of the importance of this project, in September 1973 the Army designated their five most important projects, the "Big Five" with AAH included.

Proposals were submitted by Bell, Boeing Vertol/Grumman team, Hughes, Lockheed, and Sikorsky. In July 1973, the U.S. Department of Defense selected finalists Bell and Hughes Aircraft's Toolco Aircraft Division (later Hughes Helicopters). This began the phase 1 of the competition. Each company built prototype helicopters and went through a flight test program. Hughes' Model 77/YAH-64A prototype first flew on 30 September 1975, while Bell's Model 409/YAH-63A prototype first flew on 1 October. After evaluating test results, the Army selected Hughes' YAH-64A over Bell's YAH-63A in 1976. Reasons for selecting the YAH-64A included its more damage tolerant four-blade main rotor and the instability of the YAH-63's tricycle landing gear arrangement.

The AH-64A then entered phase 2 of the AAH program. This called for building three pre-production AH-64s, and upgrading the two YAH-64A flight prototypes and the ground test unit up to the same standard. Weapons and sensor systems were integrated and tested during this time, including the new Hellfire missile.

In 1981, three pre-production AH-64As were handed over to the US Army for Operational Test II. The Army testing was successful, but afterward it was decided to upgrade to the more powerful T700-GE-701 version of engine, rated at 1,690 shp (1,259 kW). The AH-64 was named the Apache in late 1981, keeping with the Army's traditional use of American Indian tribal names for its helicopters and it was approved for full scale production in 1982. In 1983, the first production helicopter was rolled out at Hughes Helicopter's facility at Mesa, Arizona. Hughes Helicopters was purchased by McDonnell Douglas for $470 million in 1984. The helicopter unit later became part of The Boeing Company with the merger of Boeing and McDonnell Douglas in August 1997. In 1986, the incremental or flyaway cost for the AH-64A was US$7.03 million and the average unit cost was approximately US$13.9 million based on total costs.

In the mid-1980s, McDonnell Douglas studied an improved "AH-64B" design with an updated cockpit, new fire control system and other upgrades. In 1988 funding was approved for a multi-stage upgrade program to improve sensor and weapon avionic systems and incorporate some digital systems. However, improved technology was becoming available. It was decided to cancel the upgrade program for more ambitious changes.

This would lead to the more advanced AH-64D Apache Longbow. Development of the AH-64D was approved by the Defense Acquisition Board in August 1990.

The first AH-64D Apache Longbow prototype was flown on 15 April 1992, and testing of the prototypes ended in April 1995 after they had significantly outperformed the AH-64A model. On 13 October 1995 full-scale production of the Apache Longbow was approved and a $1.9 billion five year contract was signed in August 1996 to upgrade and rebuild 232 existing AH-64 Apaches.

The first production AH-64D flew on 17 March 1997 and was delivered on 31 March 1997. The cost of the AH-64D program totaled US$11 billion through 2007.

The Apache has a four-blade main rotor and a four-blade tail rotor. The crew sits in tandem, with the pilot sitting behind and above the copilot/gunner. The crew compartment and fuel tanks are armored such that the aircraft will remain flyable even after sustaining hits from 23-millimeter (0.91 in) gunfire.

The AH-64 is powered by two General Electric T700 turboshaft engines with high-mounted exhausts on either side of the fuselage. Various models of engines have been used on the Apache, those in British service use engines from Rolls-Royce instead of General Electric. In 2004, General Electric Aviation began producing more powerful T700-GE-701D engines, rated at 2,000 shp (1,500 kW) for AH-64Ds.

One of the revolutionary features at the introduction of the Apache was its helmet mounted display, the Integrated Helmet and

Display Sighting System (IHADSS); among other abilities the pilot or gunner can slave the helicopter's 30 mm automatic M230 Chain Gun to his helmet, making the gun track head movements to point at where he looks. The M230E1 can be alternatively fixed to a locked forward firing position, or controlled via the Target Acquisition and Designation System (TADS).

The AH-64 is designed to endure front-line environments and to operate during the day or night and in adverse weather using avionics, such as the Target Acquisition and Designation System, Pilot Night Vision System (TADS/PNVS), passive infrared countermeasures, GPS, and the IHADSS. A newer system that is replacing TADS/PNVS is Arrowhead (MTADS); it is manufactured by Lockheed Martin, a contract was made on 17 February 2005 to begin equipping all models of American Apaches.

The AH-64 is adaptable to numerous different roles within its context as Close Combat Attack (CCA), and has a customizable weapons loadout for the role desired. In addition to the 30-mm M230E1 Chain Gun, the Apache carries a range of external stores on its stub-wing pylons, typically a mixture of AGM-114 Hellfire anti-tank missiles, and Hydra 70 general-purpose unguided 70 mm (2.76 in) rockets. The Stinger and AIM-9 Sidewinder air-to-air missiles and the AGM-122 Sidearm anti-radiation missile were evaluated beginning in the late 1980s. The Stinger was initially selected over the AIM-9, but the US Army is considering the Starstreak air-to-air missile instead. The stub-wing pylons also have mounting points for use during ground helicopter maintenance; though in case of emergency the mount points can be used for harnessing personnel to the wings during transport. External fuel tanks can also be carried by the pylons to increase range and mission time.

The U.S. Army formally accepted its first production AH-64A in January 1984 and training of the first pilots began later that year. The first operational Apache unit, 7th Battalion, 17th Cavalry Brigade, began training on the AH-64A in April 1986 at Fort Hood, Texas. Two operational units with 68 AH-64s first deployed to Europe in September 1987 and took part in large military exercises there. The helicopter was first used in combat in 1989, during Operation Just Cause, the invasion of Panama.

The AH-64 participated in over 240 hours of combat attacking various targets, mostly at night.

During Operation Desert Storm on 17 January 1991, eight AH-64As guided by four MH-53 Pave Low IIIs, were used to destroy a portion of the Iraqi radar network. This was the first attack of Desert Storm and it allowed attack aircraft into Iraq without detection. The Apaches carried an asymmetric load of Hydra 70 flechette rockets, Hellfires, and one auxiliary fuel tank each. During the 100-hour ground war a total of 277 AH-64s took part, destroying over 500 tanks, numerous armored personnel carriers and other Iraqi vehicles. Only one AH-64 was lost in the war. It was hit by an RPG at close range and crashed, but the crew survived.

The AH-64 played roles in the Balkans during separate conflicts in Bosnia and Kosovo in the 1990s. During these deployments the Apache encountered problems such as deficiencies in training, night vision equipment, fuel tanks, and survivability. On 27 April 1999 an Apache crashed during training in Albania due to a failure with the tail rotor, causing the entire fleet in the Balkans to be grounded in December 2000. Major General Dick Cody, commanding officer of the 101st Airborne at the time, wrote a strongly worded memo to the US Army Chief of Staff about the failures in training and equipment.

The AH-64 took part in invasion of Iraq in 2003 during Operation Iraqi Freedom. In one engagement on 24 March 2003, 31 Apaches were damaged, and one Apache was shot down and captured by Iraqi troops near Karbala. The intended attack against an armored brigade of the Iraqi Republican Guard's Medina Division was unsuccessful. The tank crews had set up a "flak trap" amongst terrain and employed their guns to good effect. Iraqi officials claimed a farmer with a Brno rifle shot down the Apache, however the farmer denied involvement. The helicopter came down intact, and both the pilot and co-pilot were captured. The AH-64D was destroyed via air strike the following day.

The U.S. Apaches have been serving in Operation Enduring Freedom in Afghanistan from 2001. American AH-64Ds are flying in Iraq and Afghanistan without the Longbow Fire Control Radar as there are no armored threats to be dealt with. Most Apache helicopters that have taken heavy combat damage have been able to continue their missions and return safely. In 2006, an Apache helicopter was downed by a Soviet-made Strela 2 (SA-7) in Iraq.

The Apache is typically able to avoid hits by such missiles; however, in this instance it did not. As of 2009, 12 Apache helicopters were shot down by enemy fire during the Iraq War.

According to Boeing the U.S. Army Apache fleet has accumulated more than 2 million flight hours since the first prototype aircraft flew in 1975.

The UK operates a modified version of the Apache Longbow initially called the Westland WAH-64 Apache, and is designated Apache AH1 by the British Army.

Westland built 67 WAH-64 Apaches under license from Boeing, following a competition between the Eurocopter Tiger and the Apache for the British Army's new Attack Helicopter in 1995.

Important deviations made by AgustaWestland from the US variants of the Apache include replacing the engines with more powerful Rolls-Royce units, and the addition of a folding blade assembly for naval use; allowing British Apaches to operate from Royal Navy warships and auxiliaries.

The Eurocopter AS332 Super Puma (fig. 62) is a four-bladed, twin-engine, medium-size utility helicopter marketed for both civil and military use.

Originally designed and built by Aérospatiale, it is an enlarged and re-engined version of the original Aérospatiale Puma.

The Super Puma first flew on 13 September 1978.

Fig. 62. The Eurocopter AS332 Super Puma.

In 1974, Aérospatiale commenced development of a new medium transport helicopter based on its SA 330 Puma, announcing the project at the 1975 Paris Air Show.

While the new design was of similar layout to the AS 330, it was powered by two of the new and more powerful Turbomeca Makila turboshaft engines powering a four-bladed composite main rotor, and was designed to be withstand damage better, with a more robust fuselage structure, a new crashworthy undercarriage and the ability to withstand battle damage to the rotor blades and other key mechanical systems.

It was fitted with a ventral fin under the tail a more streamlined nose compared with the SA 330, while from the start was planned to be available with two fuselage lengths, with a short fuselage version offering similar capacity to the SA 330, which gives better performance in "hot and high" conditions and a stretched version allowing more passengers to be carried when weight is less critical.

A pre-production prototype, the SA 331, modified from a SA 330 airframe with Makila engines and a new gearbox, flew on 5 September 1977.

The first prototype of the full Super Puma made its maiden flight on 13 September 1978, being followed by a further five prototypes.

The type has proved immensely successful, chosen by 37 military forces around the world, and some 1,000 civil operators.

The Super Puma has proved especially well-suited to the North Sea oil industry, where it is used to ferry personnel and equipment to and from oil platforms.

In civilian configuration it can seat approximately 18 passengers and two crew, though since the early 2000s most oil companies have banned use of the middle-rear seat reducing effective capacity to 17+2.

This down-rating is due to difficulties encountered in evacuating through the rear-most windows in crashes at sea.

A wide variety of specialised military variants are in use, including dedicated Search and rescue (SAR) and Anti-submarine warfare (ASW) versions.

Since 1990, military Super Pumas have been marketed as the AS532 Cougar.

Fig. 63. *The Boeing CH-47 Chinook.*

The Boeing CH-47 Chinook (fig. 63) is a twin-engine, tandem rotor heavy-lift helicopter. Its top speed of 170 knots (196 mph, 315 km/h) was faster than contemporary utility and attack helicopters of the 1960s.

It is one of the few aircraft of that era, such as the C-130 Hercules and the UH-1 Iroquois, that is still in production and front line service with over 1,179 built to date. Its primary roles include troop movement, artillery emplacement and battlefield resupply. It has a wide loading ramp at the rear of the fuselage and three external-cargo hooks.

The Chinook was designed and initially produced by Boeing Vertol in the early 1960s. The helicopter is now produced by Boeing Rotorcraft Systems. Chinooks have been sold to 16 nations with the US Army and the Royal Air Force (see Boeing Chinook (UK variants)) being the largest users. The CH-47 is among the heaviest lifting Western helicopters.

In late 1956, the Department of the Army announced plans to replace the CH-37 Mojave, which was powered by piston engines, with a new, turbine-powered helicopter. Turbine engines were also a key design feature of the smaller UH-1 "Huey" utility helicopter. Following a design competition, in September 1958, a joint Army-Air Force source selection board recommended that the Army procure the Vertol medium transport helicopter. However, funding for full-scale development was not then available, and the Army vacillated on its design requirements. Some in the Army aviation corps thought that the new helicopter should be a light tactical transport aimed at taking over the missions of the old piston-engined H-21 and H-34 helicopters, and consequently capable of carrying about fifteen troops (one squad). Another faction in the Army aviation corps thought that the new helicopter should be much larger to be able to airlift a large artillery piece , and have enough internal space to carry the new MGM-31 "Pershing" Missile System.

Vertol began work on a new tandem-rotor helicopter designated Vertol Model 107 or V-107 in 1957. In June 1958, the US Army awarded a contract to Vertol for the aircraft under the YHC-1A designation. The YHC-1A had a capacity for 20 troops. Three were tested by the Army to derive engineering and operational data. However, the YHC-1A was considered by most of the Army users to be too heavy for the assault role and too light for the transport role. The decision was made to procure a heavier transport helicopter and at the same time upgrade the UH-1 "Huey" as a tactical troop transport. The YHC-1A would be improved and adopted by the Marines as the CH-46 Sea Knight in 1962. The Army then ordered the larger Model 114 under the designation HC-1B. The pre-production Boeing Vertol YCH-1B made its initial hovering flight on 21 September 1961. In 1962 the HC-1B was redesignated the CH-47A under the 1962 United States Tri-Service aircraft designation system.

The name "Chinook" alludes to the Chinook people of the Pacific Northwest. The CH-47 is powered by two turboshaft engines, mounted on each side of the helicopter's rear end and connected to the rotors by driveshafts. Initial models were fitted with engines of 2,200 horsepower. The counter-rotating rotors eliminate the need for an anti-torque vertical rotor, allowing all power to be used for lift and thrust. The ability to adjust lift in either rotor makes it less sensitive to changes in the center of gravity, important for the cargo lifting role. If one engine fails, the other can drive both rotors. The "sizing" of the Chinook was directly related to the growth of the Huey and the Army's tacticians' insistence that initial air assaults be built around the squad. The Army pushed for both the Huey and the Chinook, and this focus was responsible for the acceleration of its air mobility effort.

Improved and more powerful versions of the CH-47 have been developed since the helicopter entered service. The US Army's first major design leap was the now-common CH-47D, which entered service in 1982. Improvements from the CH-47C included upgraded engines, composite rotor blades, a redesigned cockpit to reduce pilot workload, improved and redundant electrical systems, an advanced flight control system and improved avionics. The latest mainstream generation is the CH-47F, which features several major upgrades to reduce maintenance, digitized flight controls, and is powered by two 4,733-horsepower Honeywell engines.

A commercial model of the Chinook, the Boeing-Vertol Model 234, is used worldwide for logging, construction, fighting forest fires, and supporting petroleum extraction operations. On 15 December 2006, the Columbia Helicopters company of the Salem, Oregon, metropolitan area, purchased the Type Certificate of the Model 234 from Boeing. The Chinook has also been licensed to be built by companies outside of the United States, such as Elicotteri Meridionali (now AgustaWestland) in Italy, Kawasaki in Japan, and a company in the United Kingdom.

The Army finally settled on the larger Chinook as its standard medium transport helicopter and as of February 1966, 161 aircraft had been delivered to the Army. The 1st Cavalry Division had brought their organic Chinook battalion with them when they arrived in 1965 and a separate aviation medium helicopter company, the 147th, had arrived in Vietnam on 29 November 1965. This latter company was initially placed in direct support of the 1st Infantry Division.

The most spectacular mission in Vietnam for the Chinook was the placing of artillery batteries in perilous mountain positions inaccessible by any other means, and then keeping them resupplied with large quantities of ammunition. The 1st Cavalry Division found that its Chinooks were limited to 7,000 pounds payload when operating in the mountains, but could carry an additional 1,000 pounds when operating near the coast. The early Chinook design was limited by its rotor system which did not permit full use of the installed power, and users were anxious for an improved version which would upgrade this system.

As with any new piece of equipment, the Chinook presented a major problem of "customer education". Commanders and crew chiefs had to be constantly alert that eager soldiers did not overload the temptingly large cargo compartment.

It would be some time before troops would be experts at using sling loads.

The Chinook soon proved to be such an invaluable aircraft for artillery movement and heavy logistics that it was seldom used as an assault troop carrier. Some of the Chinook fleet were used for casualty evacuation, due to the very heavy demand for the helicopters they were usually overburdened with wounded. Perhaps the most cost effective use of the Chinook was the recovery of other downed aircraft.

The Chinooks were generally armed with a single 7.62 millimeter M60 machine gun on a pintle mount on either side of the machine for self-defense, with stops fitted to keep the gunners from firing into the rotor blades. Dust filters were also added to improve engine reliability. At its peak employment in Vietnam, there were 22 Chinook units in operation.

Of the nearly 750 Chinooks in the US and Republic of Vietnam fleets, about 200 were lost in combat or wartime operational accidents. US Army supplied Chinooks to the Australian Task Force as required.

During the 1970s, the United States and Iran had a strong relationship, in which the Iranian armed forces began to use many American military aircraft, most notably the F-14 Tomcat, as part of a modernisation programme. After an agreement signed between Boeing and Elicotteri Meridionali, the Imperial Iranian Air Force purchased 20 Elicotteri Meridionali-built CH-47Cs in 1971.

The Imperial Iranian Army Aviation purchased 70 CH-47Cs from Elicotteri Meridionali during the period of 1972–1976. In late 1978, Iran placed an order for an additional 50 helicopters with Elicotteri Meridionali, but that order was canceled immediately after the revolution. Despite the arms embargo on place upon Iran, they have managed to keep their fleet operational.

In the 1978 Iranian Chinook shootdown, four Iranian CH-47C Chinooks penetrated 15–20 km into Soviet airspace in the Turkimenistan Military District. They were intercepted by a MiG-23M which shot down one Chinook, killing eight crew members, and forced a second one to land.

Chinooks were used in efforts by the Imperial Iranian loyalist forces to resist the 1979 Iranian revolution. During the war with Iraq, Iran made heavy use of its US-bought equipment, and lost at least 8 Chinooks during the 1980–1988 period; most notably during a clash on 15 July 1983, where an Iraqi Mirage F-1 destroyed three Iranian CH-47s transporting troops to the front line.

The Chinook was used both by Argentina and the United Kingdom during the Falklands War in 1982. The Argentine Air Force and the Argentine Army deployed four CH-47C (two each) which were widely used in general transport duties.

Of the Army's airframes one was destroyed on ground by a Harrier while the other was captured (and reused after the war) by the British. Both Air Force helicopters returned to Argentina and remained in service until 2002.

Approximately 163 CH-47Ds served in Kuwait and Iraq during Operations Desert Shield and Desert Storm in 1990–91.

The CH-47D has been seen wide use in Operation Enduring Freedom in Afghanistan and Operation Iraqi Freedom in Iraq.

The Chinook is being used in air assault missions, inserting troops into fire bases and later bringing food, water, and ammunition. It is also the casualty evacuation (casevac) aircraft of choice in the British Army.

In today's usage it is typically escorted by attack helicopters such as the AH-64 Apache for protection.

Its tandem rotor design and lift capacity have been found to be particularly useful in the mountainous terrain of Afghanistan where high altitudes and temperatures limit the use of the UH-60 Black Hawk. The CH-47F is being fielded by more units such as the 101st Combat Aviation Brigade and 4th Combat Aviation Brigade in the U.S. Army as it continues to operate in Afghanistan.

The Chinooks of several nations have participated in the Afghanistan War, including aircraft from Britain, Italy, the Netherlands, Spain, Canada, and Australia.

Despite the age of the Chinook, it is still in heavy demand, in part due its proven versatility and ability to operate in demanding environments such as Afghanistan.

Three CH-47 Chinooks were used to cool Reactors 3 and 4 of the Fukushima Nuclear power-plant with sea water after the 9.0 earthquake in 2011. To protect the crew from the heightened radiation levels, lead plates were attached to the floor.

The Sikorsky CH-53E Super Stallion (fig. 64) is the largest and heaviest helicopter in the United States military. It was developed from the CH-53 Sea Stallion, mainly by adding a third engine.

Sailors commonly refer to the Super Stallion as the "Hurricane Maker" because of the downwash the helicopter generates. It was built by Sikorsky Aircraft for the United States Marine Corps.

The less common MH-53E Sea Dragon fills the United States Navy's need for long range mine sweeping or Airborne Mine Countermeasures (AMCM) missions, and perform heavy-lift duties for the Navy.

The CH-53E/MH-53E are designated "S-80" by Sikorsky. Currently under development is the CH-53K, which will be equipped with new engines, new composite rotor blades, and a wider cabin.

Fig. 64. *The Sikorsky CH-53E Super Stallion.*

127

The Eurocopter Tiger (company designation EC 665), (fig. 65), is an attack helicopter manufactured by Eurocopter. In Germany it is known as the Tiger; in France and Spain it is called the Tigre.

Fig. 65. *The Eurocopter Tiger EC 665.*

The Sikorsky X2 (fig. 66) is an experimental compound helicopter with coaxial rotors under development by the American aircraft manufacturer Sikorsky Aircraft.

Fig. 66. *The Sikorsky X2.*

Sikorsky has incorporated decades of company research and development into X2 Technology helicopters. The S-69/XH-59A Advancing Blade Concept Demonstrator showed high speed was possible with a coaxial helicopter and auxiliary propulsion; the Cypher UAV expanded company knowledge of the unique aspects of flight control laws in a fly by wire aircraft with coaxial rotors; and the RAH-66 Comanche, which developed expertise in composite rotors and advanced transmission design.

On 4 May 2009, Sikorsky unveiled a mock-up of a Light Tactical Helicopter derivative of the X2.

The X2 first flew on 27 August 2008 from Schweizer Aircraft's (a division of Sikorsky Aircraft Corporation) facility at Horseheads, New York. The flight lasted 30 minutes. This began a 4-phase flight test program, to culminate with reaching a planned 250-knot top speed. The X2 completed flights with its pusher propeller fully engaged in July 2009. Sikorsky completed phase 3 of the testing with the X2 hitting 181 knots in test flight in late May 2010.

On 26 July 2010, Sikorsky announced that the X2 exceeded 225 knots (259 mph; 417 km/h) during flight testing in West Palm Beach Florida, unofficially surpassing the current FAI rotorcraft world speed record of 216 knots (249 mph) set by a modified Westland Lynx in 1986. The X2 flight was purposefully made 37 years to the date of the S-69's first flight.

On 15 September 2010, test pilot Kevin Bredenbeck achieved Sikorsky's design goal for the X2 when he flew it at a speed of 250 knots (290 mph; 460 km/h) in level flight, an unofficial speed record for a helicopter. The demonstrator also reached a speed of 260 knots (300 mph; 480 km/h) in a shallow 2° to 3° dive.

SHIPS STOVL

STOVL is an acronym for short take off and vertical landing.

This is the ability of some aircraft to take off from a short runway or take off vertically if it does not have a very heavy payload and land vertically (i.e. with no runway). The formal NATO definition (since 1991) is:

A Short Take-Off and Vertical Landing aircraft (aéronef à décollage court et atterrissage vertical) is a fixed-wing aircraft capable of clearing a 15 m (50 ft) obstacle within 450 m (1,500 ft) of commencing take-off run, and capable of landing vertically.

This is often accomplished on aircraft carriers through the use of "ski-jump" runways, instead of the conventional catapult system. STOVL use tends to allow aircraft to carry a larger payload as compared to during VTOL use, while still only requiring a short runway. The most famous example is probably the Hawker Siddeley Harrier, which though technically a VTOL aircraft, is operationally a STOVL aircraft due to the extra weight it carries at take off for fuel and armaments. The same is true of the F-35B Lightning II, which demonstrated VTOL capability in test flights but is operationally STOVL.

In 1951, the Lockheed XFV-1 and the Convair XFY tailsitters were both designed around the Allison YT40 turboprop engine driving contra-rotating propellers.

The British Hawker P.1127 took off vertically in 1960, and demonstrated conventional take off in 1961. By 1964 the first development aircraft, the Hawker Siddeley Kestrel, were flying. These were flown by a tripartite squadron of British, US and West German pilots. The first Hawker Siddeley Harrier flew in 1967.

In 1962, Lockheed built the XV-4 Hummingbird for the U.S. Army. It sought to "augment" available thrust by injecting the engine exhaust into an ejector pump in the fuselage. First flying vertically in 1963, it suffered a fatal crash in 1964. It was converted into the XV-4B Hummingbird for the U.S. Air Force as a testbed for separate, vertically mounted lift engines, similar to those used in the Yak-38 Forger. That plane flew and later crashed in 1969. The Ryan XV-5 Vertifan, which was also built for the U.S. Army at the same time as the Hummingbird, experimented with gas driven lift fans. That plane used fans in the nose and each wing, covered by doors which resembled half garbage can lids when raised. However, it crashed twice, and proved to generate a disappointing amount of lift, and was difficult to transition to horizontal flight.

Of dozens of VTOL and V/STOL designs tried from the 1950s to 1980s, only the subsonic Hawker Siddeley Harrier and Yak-38 Forger reached operational status, with the Forger being withdrawn after the fall of the Soviet Union.

Boeing had studied another odd-looking supersonic fighter in the 1960s which never made it beyond photos in Aviation Week. Rockwell International built, and then abandoned, the Rockwell XFV-12 supersonic fighter which had an unusual wing which opened up like window blinds to create an ejector pump for vertical flight. It never generated enough lift to get off the ground despite developing 20,000 lbf of thrust. The French had a nominally Mach 2 Dassault Mirage IIIV fitted with no less than 8 lift engines that flew (and crashed), but did not have enough space for fuel or payload for combat missions. The German EWR VJ 101 used swiveling engines mounted on the wingtips with fuselage mounted lift engines, and the VJ 101C X1 reached supersonic flight (Mach 1.08) on July 29, 1964. The supersonic Hawker Siddeley P.1154 which competed with the Mirage IIIV for NATO use was cancelled even as the aircraft were being built.

NASA uses the abbreviation SSTOVL for Supersonic Short Take-Off / Vertical Landing, and as of 2011, the X-35B/F-35B are the only aircraft to conform with this combination within one flight.

The experimental Mach 1.7 Yakovlev Yak-141 did not find an operational customer, but its rotating rear nozzle technology found good use with the F-35B. The F-35 Lightning II is expected to enter service by 2016.

Larger STOVL designs were considered, the Armstrong Whitworth AW.681 cargo aircraft was under development when cancelled in 1965. The Dornier Do 31 got as far as three experimental aircraft before cancellation in 1970.

Although mostly a VTOL design, the V-22 Osprey has increased payload when taking off from a short runway.

The Hawker Siddeley Harrier, colloquially the "Harrier Jump Jet", was developed in the 1960s and was the first generation of the Harrier series of aircraft. It was the first operational close-support and reconnaissance fighter aircraft with Vertical/Short Takeoff and Landing (V/STOL) capabilities and the only truly successful V/STOL design of the many that arose in that era. The Harrier was produced directly from the Hawker Siddeley Kestrel prototypes following the cancellation of a more advanced supersonic aircraft, the Hawker Siddeley P.1154. The Royal Air Force (RAF) ordered the Harrier GR.1 and GR.3 (fig. 67) variants in the late 1960s. It was exported to the United States as the AV-8A, for use by the US Marine Corps (USMC), in the 1970s.

Fig. 67. *The Hawker Siddeley Harrier; this is an RAF Harrier GR.3.*

The RAF positioned the bulk of their Harriers in West Germany to defend against a potential invasion of Western Europe by the Soviet Union; the unique abilities of the Harrier allowed the RAF to disperse their forces away from vulnerable and well-known airbases. The USMC used their Harriers primarily for close air support, operating from amphibious assault ships, and if needed forward operating bases. Harrier squadrons saw several deployments overseas. The Harrier's ability to operate with minimal ground facilities and very short runways allowed it to be used at locations unavailable to other fixed-wing aircraft.

In the 1970s the British Aerospace Sea Harrier was developed from the Harrier for use by the Royal Navy (RN) on Invincible class aircraft carriers.

The Sea Harrier and the Harrier were crucial during the 1982 Falklands War, in which the aircraft proved to be flexible and versatile. The RN Sea Harriers provided fixed-wing air defence while the RAF Harriers focused on ground-attack missions in support of the advancing British land force.

The Harrier was also extensively redesigned as the AV-8B Harrier II and British Aerospace Harrier II by the team of McDonnell Douglas and British Aerospace.

The innovative Harrier family and its Rolls-Royce Pegasus engines with thrust vectoring have generated long-term interest in V/STOL aircraft. Similar V/STOL operational aircraft include the contemporary Soviet Yakovlev Yak-38 as well as one variant of the Lockheed Martin F-35 Lightning II, which is currently under development.

The F-35 Lightning II joint strike fighter (JSF), is being developed by Lockheed Martin Aeronautics Company for the US Air Force, Navy and Marine Corps and the UK Royal Navy.

The stealthy, supersonic multirole fighter was designated the F-35 Lightning II in July 2006. The JSF is being built in three variants: a conventional take-off and landing aircraft (CTOL) for the US Air Force; a carrier variant (CV) for the US Navy; and a short take-off and vertical landing (STOVL) aircraft for the US Marine Corps and the Royal Navy (fig. 68).

Fig. 68. *The F-35B short take-off and vertical landing (STOVL) variant for the US Marine Corps and the Royal Navy.*

The Lockheed Martin F-35 Joint Strike Fighter short takeoff/vertical landing (STOVL) variant flew faster than the speed of sound for the first time June 10, achieving a significant milestone. The aircraft accelerated to Mach 1.07 (727 miles per hour) on the first in a long series of planned supersonic flights.

"For the first time in military aviation history, supersonic, radar-evading stealth comes with short takeoff/vertical landing capability," said Bob Price, Lockheed Martin's F-35 U.S. Marine Corps program manager. "The supersonic F-35B can deploy from small ships and austere bases near front-line combat zones, greatly enhancing combat air support with higher sortie-generation rates." The F-35B will enter service for the Marines, the United Kingdom's Royal Air Force and Royal Navy, and the Italian Air Force and Navy.

The F-35B is the short takeoff and vertical landing (STOVL) variant of the aircraft. Similar in size to the A variant, the B sacrifices about a third of the other versions fuel volume to make room for the vertical flight system. Takeoffs and landing with vertical flight systems are by far the riskiest, and in the end, a decisive factor in design. Like the AV-8B Harrier II, the B's guns will be carried in a ventral pod. Whereas the F-35A is stressed to 9 g, the F-35B is stressed to 7 g. The F-35B was unveiled at Lockheed Martin's Fort Worth plant on 18 December 2007, and the first test flight was on 11 June 2008.

The three-bearing swivel nozzle that directs the full thrust of the afterburning jet engine is moved by a "fueldraulic" actuator, using pressurized jet fuel.

Unlike the other variants, because it can land vertically the F-35B has no landing hook. The "STOVL/HOOK" button in the cockpit initiates conversion instead of dropping the hook. The F-35B sends jet thrust directly downwards during vertical takeoffs and landing and the nozzle is being redesigned to spread the output out in an oval rather than a small circle so as to limit damage to asphalt and ship decks.

The United States Marine Corps plans to purchase 340 F-35Bs, to replace all current inventories of the F/A-18 Hornet (A, B, C and D-models), and AV-8B Harrier II in the fighter, and attack roles.

The Royal Air Force and Royal Navy had planned to use the F-35B to replace their Harrier GR9s.

One of the Royal Navy requirements was that the F-35B design should have a Shipborne Rolling and Vertical Landing (SRVL) mode so that wing lift could be added to powered lift to increase the maximum landing weight of carried weapons.

This method of landing is slower than wire arrested landing and could disrupt regular carrier operations, as the landing method uses the same pattern of approach as wire arrested. With SRVL, the aircraft is able to "bring back" 2 × 1K JDAM, 2 × AIM-120 and reserve fuel. However, in October 2010, Prime Minister David Cameron announced that the UK would change their F-35 order to the CATOBAR F-35C variant.

Commandant of the Marine Corps, General James Amos has said that, in spite of its increasing costs and schedule delays, there is no plan B to substitute for the F-35B.

The F-35B is larger than the aircraft it replaces, which required the USS America (LHA-6) to be designed without needed well deck capabilities.

In 2011, the USMC and USN signed an agreement that the USMC will purchase 340 F-35B and 80 F-35C while the USN will purchase 260 F-35C.

The five squadrons of Marine Corps F-35Cs will be assigned to the Navy carriers while the Marine Corps F-35Bs will be used on Amphibious ships and ashore.

On 6 January 2011, Gates said that the 2012 budget would call for a two year pause in F-35B production during which the aircraft may be redesigned, or canceled if unsuccessful.

Gates stated, "If we cannot fix this variant during this time frame, and get it back on track in terms of performance, cost and schedule, then I believe it should be canceled."

Lockheed Martin executive vice president Tom Burbage and former Pentagon director of operational testing Tom Christie have said that most of the delays in the total program have been due to issues with the F-35B, which forced massive redesigns on the other versions.

The USMC intends to declare Initial Operational Capability with about 50 F-35s running interim Block 2B software in the 2014 to 2015 timeframe.

The Bell-Boeing V-22 Osprey (fig. 69) is an American multi-mission, military, tiltrotor aircraft with both a vertical takeoff and landing (VTOL), and short takeoff and landing (STOL) capability.

It is designed to combine the functionality of a conventional helicopter with the long-range, high-speed cruise performance of a turboprop aircraft.

The V-22 originated from the United States Department of Defense Joint-service Vertical take-off/landing Experimental (JVX) aircraft program started in 1981.

The team of Bell Helicopter, and Boeing Helicopters was awarded a development contract in 1983 for the tiltrotor aircraft. The Bell Boeing team jointly produce the aircraft.

The V-22 first flew in 1989, and began flight testing and design alterations; the complexity and difficulties of being the first tiltrotor intended for military service in the world led to many years of development.

Fig. 69. *The Bell-Boeing V-22 Osprey.*

The United States Marine Corps began crew training for the Osprey in 2000, and fielded it in 2007; it is supplementing and will eventually replace their CH-46 Sea Knights.

The Osprey's other operator, the U.S. Air Force fielded their version of the tiltrotor in 2009. Since entering service with the U.S. Marine Corps and Air Force, the Osprey has been deployed in both combat and rescue operations over Iraq, Afghanistan and Libya (fig. 70).

Fig. 70. The Bell-Boeing MV-22B Osprey, (M from Marine).

The PA-23 was the first twin-engine design from Piper and was developed from a proposed "Twin Stinson" design inherited when Piper bought the Stinson Division of the Consolidated Vultee Aircraft Corporation (fig. 71). The prototype PA-23 was a four-seater low-wing all-metal monoplane with a twin tail, powered by a two 125 hp Lycoming O-290-D piston engines the prototype first flew 2 March 1952. The aircraft performed badly and it was redesigned with a single vertical stabilizer and an all-metal rear fuselage and more powerful 150 hp Lycoming O-320-A engines. Two new prototype of re-designed aircraft now named Apache were built in 1953 and entered production in 1954; 1,231 were built. In 1958, the Apache 160 was produced by upgrading the engines to 160 hp (119 kW), and 816 were built before being superseded by the Apache 235, which went to 235 hp (175 kW) engines and swept tail surfaces (119 built).

In 1958 an upgraded version with 250 hp (186 kW) Lycoming O-540 engines and adding a swept vertical tail was produced as the PA-23-250 and was named Aztec.

These first models came in a five-seat configuration which became available in 1959. In 1961 a longer nosed variant the Aztec B entered production. The later models of the Aztec were equipped with IO-540 fuel-injected engines and six-seat capacity, and continued in production until 1982. There were also turbocharged versions of the later models, which were able to fly at higher altitudes.

The US Navy acquired 20 Aztecs, designating them UO-1, which changed to U-11A when unified designations were adopted in 1962.

In 1974, Piper produced a single experimental PA-41P Pressurized Aztec concept.

This concept was short-lived, however, as the aspects of the Aztec that made it so popular for its spacious interior and ability to haul large loads did not lend themselves well to supporting the sealed pressure vessel required for a pressurized aircraft. The project was scrapped, and the one pressurized Aztec produced, N9941P, was donated to Mississippi State University, where it was used for testing purposes.

In 2000, N9941P was donated to the Piper Aviation Museum in Lock Haven, PA, on the condition that it never be flown again. It now sits there on display.

Fig. 71. *The Piper PA-23-250 Aztec F aircraft.*

The DHC-6 Twin Otter is a Canadian 19-passenger STOL (Short Takeoff and Landing) utility aircraft developed by de Havilland Canada and currently produced by Viking Air. The aircraft's fixed tricycle undercarriage, STOL abilities and high rate of climb have made it a successful cargo, regional passenger airliner and MEDEVAC aircraft. In addition, the Twin Otter has been popular with commercial skydiving operations, and it is used by the United States Army Parachute Team (fig. 72).

Development of the aircraft began in 1964, with the first flight on May 20, 1965. A twin-engined replacement for the single-engined Otter had been planned by de Havilland Canada. Twin engines not only provided improved safety but also allowed for an increase in payload while retaining the renowned STOL qualities. Design features included double slotted trailing edge flaps and ailerons that work in unison with the flaps to boost STOL performance. The availability of the 550 shp (410 kW) Pratt and Whitney Canada PT6A-20 propeller turbine engine in the early 1960s made the concept of a twin more feasible. To bush operators, the improved reliability of turboprop power and the improved performance of a twin-engined configuration made it an immediately popular alternative to the single engine, piston-powered Otter which had been flying since 1951.

The first six aircraft produced were designated Series 1, indicating that they were prototype aircraft. The initial production run consisted of Series 100 aircraft, serial number seven to 115 inclusive. In 1968, Series 200 production began with serial number 116.

Changes made at the beginning of Series 200 production included improving the STOL performance, adding a longer nose that was equipped with a larger baggage compartment (except to aircraft fitted with floats) and fitting a larger door to the rear baggage compartment. All Series 1, 100 and 200 aircraft and their variants (110, 210) were fitted with the 550 shaft horsepower PT6A-20 engines.

In 1969, the Series 300 was introduced, beginning with serial number 231. Both aircraft performance and payload were improved by fitting more powerful PT6A-27 engines. This was a 680 hp (510 kW) engine that was flat-rated to 620 hp (460 kW) for use in the Series 300 Twin Otter. The Series 300 proved to be the most successful variant by far, with 614 Series 300 aircraft and their sub-variants (Series 310 for United Kingdom operators, Series 320 for Australian operators, etc.) sold before production ended in 1988.

Fig. 72. *De Havilland Canada DHC-6-100 Twin Otter aircraft.*

After Series 300 production ended, the remaining tooling was purchased by Viking Air of Victoria, British Columbia, who manufacture replacement parts for all of the out of production de Havilland Canada aircraft. On February 24, 2006 Viking purchased the type certificates from Bombardier Aerospace for all the out of production de Havilland DHC-1 through DHC-7 aircraft. The ownership of the certificates gives Viking the exclusive right to manufacture new aircraft.

On July 17, 2006, at the Farnborough Air Show, Viking Air announced its intention to offer a Series 400 Twin Otter. On April 2,

2007 Viking announced that with 27 orders and options in hand, it was restarting production of the Twin Otter, equipped with a more powerful Pratt & Whitney Canada PT6A-34/35 engine. As of November 2007, 40 firm orders and 10 options had been taken and a new assembly plant established in Calgary, Alberta. Zimex Aviation of Switzerland received the first new production aircraft, serial number 845, in July 2010.

Major changes introduced with the Series 400 include Honeywell Primus Apex fully integrated avionics, deletion of the AC electrical system, deletion of the beta backup system, modernization of the electrical and lighting system, and use of composites for non-load-bearing structures such as doors.

The Edgley EA-7 Optica (fig. 73) was a British light aircraft designed for observation work, intended as a low-cost alternative to helicopters, retailing originally at around US$200,000.

The Optica, designed by John Edgley and built by Brooklands Aerospace, had an unusual configuration with a fully-glazed forward cabin seating three across, reminiscent of an Alouette helicopter. Behind it was situated a Lycoming flat-six engine powering a ducted fan, twin boom cantilever tailplane with twin rudders and a high-mounted single elevator. The fixed tricycle undercarriage had the nosewheel offset to the left. The wings were unswept and untapered, and the aircraft was of a fairly standard all-metal construction with stressed aluminium skin. The aircraft's distinctive appearance led to it being known as the "bug-eye" in some popular reports.

It first flew on 14 December 1979, powered by a 150 hp (112 kW) Lycoming O-320 engine and flown by the chief pilot of the Cranfield College of Aeronautics. The Optica, now powered by a more powerful Lycoming O-540, entered production in 1983, achieving certification on 8 February 1985. A crash of police Optica G-KATY on 15 May 1985 killed two members of the Hampshire Constabulary. The cause was suspected to be a stall: insufficient airspeed during a turn causing instability. The reason for the low speed was never established. This led to the bankruptcy of Edgley, with Optica Industries being formed in October 1985 to continue production and 25 were built before a fire caused by arson destroyed the factory and all but one flying Optica. The company was reformed again as Brooklands Aircraft, and the Optica returned to production, production ceasing in March 1990, when Brooklands Aircraft went bankrupt. The Design of the Optica has now been bought by John Edgley once more (along with the design for the FLS Sprint 160). Edgley hopes to put both types into production and further to that goal the Optica 300 Series s/n 021 G-BOPO is being restored as a UK type demonstrator.

Fig. 73. Edgley EA-7 Optica aircraft.

The P-791 is an experimental aerostatic/aerodynamic hybrid airship developed by Lockheed-Martin corporation (fig. 74). The first flight of the P-791 was on 31 January 2006 at the company's flight test facility on the Palmdale Air Force Plant 42. It has a unique tri-hull shape, with disk-shaped cushions on the bottom for landing. A very similar design can be seen in the Long-Endurance Multi-intelligence Vehicle (LEMV).

Fig. 74. Lockheed Martin P-791 aircraft.

The P-791 is an example of a hybrid airship. In such designs, part of the weight of the craft and its payload are supported by aerostatic (buoyant) lift and the remainder is supported by aerodynamic lift.

The combination of aerodynamic and aerostatic lift is an attempt to benefit from both the high speed of aerodynamic craft and the lifting capacity of aerostatic craft. Critics of the hybrid approach have labeled it as being the "worst of both worlds" in that such craft require a runway for take-off and landing, are difficult to control and protect on the ground, and have relatively poor aerodynamic performance.

Proponents of hybrid designs claim that these shortcomings can be overcome through advanced technologies. In particular, it has been proposed that buoyancy control mechanisms can minimize or eliminate the need for a runway. Recently, Worldwide Aeros Corporation has successfully tested its COSH system which will regulate its upcoming ML866's static heaviness.

No other hybrid airship design has been developed past the initial experimental stages. Although many such designs have been proposed, very few have flown. One hybrid airship design that did take flight was the Aereon 26. The development of this aircraft was documented in the book "The Deltoid Pumpkinseed" by John McPhee.

Although Lockheed-Martin is developing a design for the DARPA project WALRUS, the company claims that the P-791 is unrelated to WALRUS. Nonetheless, the design represents an approach that may well be applicable to WALRUS. Some believe that Lockheed-Martin has used the secret P-791 program as a way to get a head-start on Worldwide Aeros Corp, the other Phase I WALRUS competitor.

The company has released no details about the design of the aircraft. However, from a distance, the P-791 appears to be essentially identical in design to the SkyCat design unsuccessfully promoted by the now defunct British company Advanced Technologies Group (ATG). SkyCat technologies are once again being developed and promoted in the US by the Hybrid Aircraft Corporation (HAC) located in New Mexico. Press reports have also confirmed that the P-791 incorporates some of the most distinctive design features of the SkyCat, specifically the use of hover/suction skirts as "landing gear."

Industry observers have noted that engineering individuals closely associated with HAC later worked on the WALRUS project and P-791 at Lockheed-Martin.

A plane of high overall gauge is Airbus A300B4-608ST Super Transporter (fig. 75).

Airbus A300B4-608ST Super Transporter aircraft

Fig. 75. Airbus A300B4-608ST Super Transporter aircraft.

It can carry even and an aircraft (fig. 76).

Fig. 76. *Airbus A300B4-608ST Super Transporter aircraft.*

The BD-5 Micro (fig. 77) is a series of small, single-seat homebuilt aircraft created in the late 1960s by US aircraft designer Jim Bede and introduced to the market primarily in "kit" form by the now-defunct Bede Aircraft Corporation in the early 1970s.

The BD-5 has a small, streamlined fuselage holding its semi-reclined pilot under a large canopy, with the engine installed in a compartment in the middle of the fuselage, and a propeller or jet engine in the BD-5J variant, mounted immediately to the rear of the cockpit. The combination of fighter-like looks and relatively low cost led to the BD-5 selling over 5,000 kits or plans, with approximately 12,000 orders being taken for a proposed factory-built FAA certified version. However, few of the kit versions were actually completed due to the company's bankruptcy in the mid-1970s, and none of the factory built "D" models produced, brought on by the failure to deliver a reliable engine for the design.

In total, only a few hundred BD-5 kits were completed, although many of these are still being flown today. The BD-5J version holds the record for the world's lightest jet aircraft, weighing only 358.8 lb (162.7 kg).

Fig. 77. Bede BD-5B aircraft.

Development of the "Micro" dates back as early as 1967, when Jim Bede was inspired by the Schleicher ASW 15. At the time, however, Bede was working on the Bede BD-4 design. The BD-4 was a fairly conventional looking high-wing four-seater, but it offered good performance and was fairly inexpensive. Over the lifetime of the company about 600 BD-4s were sold, a success by any measure.

Serious work on the Micro started in 1970, with construction of the prototype starting in earnest late that year. While the BD-4 was fairly conventional looking, the Micro was a radical design. It is an extremely small one-seat design that looked more like a jet fighter than a "prop plane," with the pilot sitting in a semi-reclined position under a large fighter-like plexiglas canopy only inches above the pilot's head. Behind the cockpit was a compartment housing a two-cylinder air-cooled 40 hp engine driving a pusher propeller. Two versions were planned, the BD-5A with "short" 14' 3" (4.34 m) wings tuned for high speeds and acrobatics, and the BD-5B with a 21' 6" (6.55 m) wings for longer range and powered glider use. Builders could optionally buy both wings, switching them in about 10 minutes.

For improved performance the aircraft featured both a V-tail and retractable landing gear in order to reduce drag. Calculated drag was so low that spoilers were added to the wing in order to improve deceleration for landing. This was apparently the first application of spoilers on a light aircraft. The low drag implied excellent performance; with the short wings it would reach 210 mph (340 km/h) in cruise, while the long-wing BD-5B would be only slightly slower and have an extended range of 1,215 miles.

In addition to being easy to fly, the BD-5 was also intended to be easy to build and own. The fuselage was constructed primarily from fiberglass panels over an aluminum frame, reducing construction time to only a few hundred hours. Although the early designs required some welding in the landing gear area, it was planned that this would be removed in the kit versions, so construction would require no special tooling or skills. Even the cost of operation would be extremely low, offering fuel efficiency of almost 40 mpg. With the wings removed the aircraft could be packed into a small custom trailer, allowing it to be towed away by car for storage in a garage, and from there to any suitable flat area for takeoff.

Bede published an information booklet about the BD-5 in November 1970. Several very positive magazine articles appeared at this point. The October 1971 issue of Science & Mechanics had the BD-5 on the cover, listing the price as $1,950 and a top speed of 215 MPH. The associated article showed the construction of the original prototype, with numerous claims about how easy it was to construct. The August 1973 issue of Popular Science also covered the aircraft, although it listed the price at $2,965. A feeding frenzy followed as the "mini fighter" generated intense demand. As one author put it, "Even before the plane first left the ground, thoughts of flying the sleek, bullet-shaped aircraft with its pusher prop stimulated the imagination of nearly everyone who had heard of the program."

On February 24, 1971, the first $200 deposit to reserve a "place in line" to receive a kit was accepted, with the target shipping date being May 24, 1972. By August 1971, 800 deposits had been taken, even though the first BD-5 prototype had yet to complete high-speed taxi tests. By the end of the year, they had over 4,300 orders, making it one of the most popular general aircraft projects in modern history.

The prototype, N500BD, flew briefly on September 12, 1971, powered by a 36 hp Polaris Industries snowmobile engine. The stability of the aircraft with the original V-tail was marginal at best, and clearly needed a redesign. With the original fiberglass fuselage this was a time consuming process, so the decision was made to switch to an all-metal fuselage with the components incorporating compound curves produced using hydroformed aircraft-grade aluminum alloy. These could be modified with relative ease during the testing cycle. It also made economic sense as the orders rolled in, as assembly line production of stamped metal parts is expensive to set up but less expensive in the long run.

By December 1971 the tooling for the new fuselage was in development. The aircraft now featured a longer, more pointed nose, whereas the more ovoid N500BD had been patterned on the ASW 15. While this work was in progress, Bede continued to experiment with modifications to the tail, eventually abandoning the V-tail and changing to a more conventional vertical rudder and horizontal elevator layout with highly swept surfaces. Further testing on N500BD showed flow interference between the horizontal surfaces and the propeller, and the elevator was raised six inches to correct it, placing it about mid-way up the rear fuselage.

The first example of the new fuselage arrived in March 1972, and was fitted with a new Kiekhaefer Aeromarine engine Bede had seen at the Oshkosh Airshow in 1971. Finished as N501BD, numerous small delays prevented it from flying until July 11, 1972. These flights demonstrated continued problems with the tail design, which was again redesigned, losing the sweep and becoming much more conventional.

The program was now far too large for Bede to handle alone. In March 1972, he hired Burt Rutan to head the flight test department, who was soon followed by Les Berven as chief test pilot. They took over development, giving Bede more time to work on business issues. This was proving difficult enough, as Kiekhaefer and Bede could not reach an agreement about deliveries, forcing him to change to a similar 440 cc 40 hp Hirth Motoren design, but then selecting a larger 650 cc 55 hp Hirth engine instead.

Several additional problems turned up during testing. Stick forces were very low, but this was easily addressed by making the servo tabs larger. A more worrying development was that the engines all had problems with mixture due to changes in engine speed or load, which led to rough engine operation. In August Bede was demonstrating the BD-5 to the FAA in order to receive permission to fly at Oshkosh, when the engine seized. On its deadstick landing, the aircraft overran the runway, buckling the nose gear. The air-fuel mixture was identified as the cause of the crash of N501BD in September 1972 when the mixture control broke and Berven had to execute a forced landing. Since N502BD would be ready in two months, N501BD was not repaired.

However, N502BD ran into problems of its own. The earlier models used a variable speed belt drive system to transfer power from the engine to the propeller shaft, but this was removed from N502BD and it suddenly began exhibiting a serious vibration problem. Experts were called in, and additional bearings corrected the problem, but it was not until March 26, 1973 that N502BD flew. From then on the test program seemed to go more smoothly.

By the time the test program neared its conclusion the aircraft had undergone major changes. One victim of the program was the shorter "A" wing, which calculations showed would only improve performance at speeds very close to Vmax (the highest available speed). Flight testing also showed the landing speed with the smaller wing was decidedly fast. Split flaps and spoilers had also disappeared. The canopy and cockpit dimensions had changed and the aircraft had new landing gear systems and a completely new tail section. More ominous was the fact that the engines had already been changed twice. What remained, however, was the basic concept of the fighter-like pusher aircraft which, if anything, had improved in looks.

By this point it seemed the basic design was complete, and Bede turned his attention to other projects. One was a jet-powered BD-5, the BD-5J, which is detailed below. Another was the Bede BD-6, a single-seat version of the BD-4 based on the same Hirth engine being used in the BD-5. Still another was the "new" Bede BD-7, a two-seat side-by-side version of the BD-5 of which a prototype was built. There was even an attempt to sidestep the engine problem with the BD-5S, a glider (S for Sailplane) version with lengthened wings and no engine, which prompted Air Progress magazine to sarcastically note, "At last, a BD-5 with no engine problems". This glider version did not fly well and the project was scrapped. Bede also decided to seek FAA certification of the BD-5D as a production aircraft and sell it complete, and began taking $600 deposits for this model.

By the middle of 1973 the basic design was complete and the tooling set up for production. The engines were the only part holding up deliveries, so Bede offered to ship the kit with the engine to follow. This was a fairly attractive option; it meant the builder could "get to work" and hopefully complete the airframe by the time the engine arrived, at that point expected in September 1973. Many builders took the company up on the offer, only to receive incomplete kits and plans.

All three Hirth engines were offered; builders could keep the 40 hp design, or "trade up" to the larger 55 hp or 70 hp engines. The latter, which Bede had developed with Hirth, was now considered the baseline engine for the aircraft as the original 40 hp proved to be of insufficient power. In a late 1973 newsletter to prospective owners, Bede suggested the 70 hp model and discouraged use of the smaller engines. Prices had risen throughout the 30 months since the deposits were first taken. Originally priced at $1,799, the base price was raised to $2,599 with the 55 hp, and owners were offered a "trade up" for the difference in price if they had ordered the aircraft with the original 40 hp engine.

When 1974 came around, the engines were still not being delivered in quantity, although some started to arrive early that year. At that point, unexpectedly, Hirth went bankrupt after about 500 of the engines had shipped. Once again the design lacked a suitable engine, but this time the search for a replacement ended with a Xenoah design from Japan. Development of this engine was lengthy, and in the end it would not be certified for export until 1978, although this was not expected at the time.

In the meantime, Bede came up with another novel solution to the problems of converting pilots to the new aircraft. They took an engine-less example and bolted it to the front of a pickup truck on a trapeze, attaching the pilot's throttle control to the truck's. Pilots could test fly the aircraft without danger - if a problem developed the driver of the truck simply hit the brakes.

After more than 5,100 kits had been delivered to prospective builders, the kits stopped shipping as well. Although the company was effectively bankrupt at this point, work on the BD-5D continued for some time. The bankruptcy became official in 1979, by which point the BD-5 project was long dead. During the bankruptcy proceedings it was learned that the money ostensibly being used to build kits was instead being spent on a variety of projects. As a result, Bede entered a consent decree with the FTC to no longer accept deposits on aircraft for a period of 10 years.

Many owners stored, abandoned, or sold their incomplete kits, but a few hundred diehard builders finished them with a variety of engine solutions designed by third parties and former Bede Aircraft dealers. Having to hunt for an engine was only one problem. The time to build the aircraft was much longer than quoted, as much as 3,000 to 3,500 hours. Some of this was due to the need to fit their selected engine into an airframe designed for the Hirth, which was no longer available. Additionally, some of the kits were shipped with missing parts, adding to the confusion. All of this led to a rash of kits being sold for fire sale prices, although this did allow the builders that were looking to complete their kits to do so at bargain prices. While Bede claimed the aircraft could be put together by anyone in a garage, builders generally agree that doing so without proper construction techniques could result in a potentially dangerous aircraft. One way to overcome that issue is to use a set of properly laid-out jigs to align and drill the pilot holes for the airframe, wings and other components. For all of these reasons, it was some time before completed BD-5s started to appear.

Over the next few years the aircraft garnered a terrible safety record. Although Bede had suggested using the B wings, the earliest kits shipped only with the short "A" wings. All four examples completed with these wings crashed on their first flight, three on takeoff and one lasted long enough to crash on landing, three of the four causing fatalities. Of the first 25 completed, with both the "A" and "B" wings, 14 crashed with 9 fatalities.

Even when examples with the "B" wings were completed, the safety record did not improve greatly. Several crashes in the -5B models were found to have taken place due to engine failure on takeoff, both due to the mix of "oddball" engines as well as endemic cooling problems. The reason this is such an issue with the BD-5 is twofold – the high line of thrust means an engine failure immediately results in an unexpected (for most pilots) nose-up attitude change. Pilots who fail to fly the aircraft first and then attempt to restart the engine inevitably stall, with the associated consequences. This was aggravated by the fact the original wing had a very sharp stall with little warning and a nasty tendency to "snap roll." To make matters worse, a documented manufacturing error in some wing skins delivered to kit builders exacerbated the problem. A rather small center of gravity range also added to the problems of properly trimming the aircraft.

With the demise of the Bede Aircraft Company, the BD-5 entered a sort of limbo while builders completed their kits. The early safety problems and the challenge of adapting a suitable engine exacerbated delays. Over the next few years, however, solutions to most of these problems arrived in one form or another. Many other changes have also been incorporated to improve the original design. Today the BD-5 is a rewarding, if demanding aircraft.

For instance, the problem of finding a suitable engine with 60 to 70 hp yet still weighing under 100 lbs was a serious problem in the 1970s, but today there are a number of "off the shelf" designs in this class. The widely available Rotax 582 is a 65 hp engine of 80 lbs in standard configuration, almost tailor-made for the BD-5. Other engines successfully used in BD-5s include the Subaru EA-81, Honda EB-1 and EB-2 (with and without turbocharging), Hirth 2706, AMW 225-3 and 2SI 808. The current record holder of the FAI C-1a/0 (300 kg or less takeoff weight) class speed record over a 3 km (1.9 mi) course at restricted altitude is a BD-5A (listed as BD-5B but used -5A wings for the record attempt) with a Rotax 618UL 74 hp two-stroke, two-cylinder water-cooled engine.

Problems with the abrupt stall were mostly addressed by Harry Riblett, an airfoil designer who documented a procedure to apply a slight reprofile of the wing root airfoil which softened the stall response of the aircraft without any significant performance degradation. The reprofile presents other unique problems associated with the way it is applied to the wing upper surface, essentially gluing foam to the aluminum skin and covering with fiberglass. Similarly, the small center-of-gravity range has since been addressed with 5.5- to 13-inch stretch kits for the fuselage.

Several companies were formed to help builders complete their kits, and many of the aftermarket modifications were worked into these services. Today, BD Micro Technologies of Siletz, Oregon continues to offer kit building support, including new-build kits featuring (optionally) all of these modifications, and even the BD-5T, a turboprop version of the BD-5 using a modified Solar T62 turbine powering a mechanically-controlled variable-pitch propeller. Alturair, Inc. of San Diego, California also offers extensive parts and construction assistance services.

An unusual adaptation of the BD-5, the Acapella 100, appeared in the early 1980s. Designer Carl D. Barlow of Option Air

Reno mated a BD-5 fuselage with a distinctive twin-boom empennage and fitted it with a 100 hp Continental O-200 engine.

Later, a 200 hp Lycoming IO-360 was fitted, and the wings shortened from 26.5 feet to 19.5 feet, becoming the Acapella 200-S model. The prototype of this aircraft was first flown on June 6, 1980, with pilot Bill Skiliar at the controls.

Nonetheless, it flew poorly and was difficult to control. Only the one prototype was built and it was donated to the Experimental Aircraft Association's Airventure Museum in Oshkosh, Wisconsin, USA, where it is occasionally placed on display.

Bede Aircraft Company has since re-formed and has been working on several new designs.

Bede has hinted at a two-seat tandem version of the aircraft called the "Super BD-5" using a certified aircraft engine and a number of modifications and improvements, but to date nothing other than a preliminary design drawing has been made available.

While the new Hirth engine was being tested, Bede decided to create an unconventional variant of the BD-5 with a small jet engine. The result was the sleek BD-5J, a 300 mph (480 km/h) aircraft.

The design used the Sermel TRS-18-046 turbojet (now Microturbo, a division of Turbomeca), which produced 225 lbs of thrust and was used on a Caproni certified motorglider design.

The original engines were produced under license by Ames Industrial in the USA.

The wing was modified to an "intermediate" size between the original A and B wings, with a 17 ft span.

Bob Bishop had purchased 20 BD-5J kits as soon as they had appeared, and many of the flying examples started life in this batch of twenty.

Versions from the original batch became a popular airshow fixture. Throughout the 1980s and until 1991, Coors flew two of them as the "Silver Bullets." Budweiser also had a BD-5J called the "Bud Light Jet", but that contract has long expired and the aircraft was lost as a result of an engine compartment fire from which Bob Bishop successfully bailed out.

The aircraft also appeared in the opening sequence of the James Bond film, Octopussy.

Many of these aircraft have since been involved in crashes. The loss of the Bud Light Jet was caused by an incorrectly specified fuel flow sending unit which burst in mid-flight and caused fuel to be sprayed directly into the engine compartment.

The fuel ignited when it came in contact with the hot components of the turbojet engine, forcing the pilot to trade speed for altitude, climb and bail out.

The aircraft then went into a flat spin and pancaked into the ground, but was sufficiently intact to allow the cause of the fire to be determined relatively quickly.

On June 16, 2006, while practicing for an air show at Carp Airport in Ottawa, Canada, Scott Manning fatally crashed in his "Stinger Jet," the last BD-5J that remained on the airshow circuit.

The Transportation Safety Board of Canada investigated the accident and issued a report assigning the probable cause to the incorrect installation of the right wing, which caused the flap on that wing to suddenly retract in flight and create a "split flap" condition. The aircraft rolled to the right and Manning was unable to recover in time.

Recently, the BD-5J operates in the national security arena. The aircraft is certified by the United States Department of Defense as a cruise missile surrogate, with Bishop's Aerial Productions offering a version known as the Smart-1 (Small Manned Aerial Radar Target, Model 1). The radar return and general performance characteristics make it a useful aid in training.

On June 27, 2006, while flying one of these aircraft, pilot Chuck Lischer, a highly experienced professional air show pilot, impacted trees on final approach to the Ocean City Municipal Airport in Ocean City, Maryland in a fatal accident.

The BD-5J has also held the Guinness record for the World's Smallest Jet for more than 25 years.

Bob Bishop originally garnered the record with one of his jets, and in November 2004 the record changed hands to Juan Jiménez,who purchased the aircraft from the original builder whose BD-5J weighed in at 358.8 lb (162.7 kg) empty weight, 80 lb (36 kg) lighter than Bishop's jet and the lightest documented weight for a BD-5.

The jet has not yet flown due to significant mechanical issues, turbine starting and safety concerns.

Another monster is Boeing 707-385C Phalcon aircraft (fig. 78).

Fig. 78. Boeing 707-385C Phalcon aircraft.

INVISIBLE AIRCRAFT

Stealth aircraft are aircraft that use stealth technology to avoid detection by employing a combination of features to interfere with radar as well as reduce visibility in the infrared, visual, audio, and radio frequency (RF) spectrum. Development of stealth technology likely began in Germany during World War II. Well-known modern examples of stealth aircraft include the United States' F-117

Nighthawk (1981–2008, fig. 79), the B-2 Spirit (fig. 80), the F-22 Raptor (fig. 81), and the F-35 Lightning II (fig. 82).

Fig. 79. *A F-117 Nighthawk stealth strike aircraft.*

While no aircraft is totally invisible to radar, stealth aircraft prevent conventional radar from detecting or tracking the aircraft effectively, reducing the odds of a successful attack. Stealth is the combination of passive low observable (LO) features and active emitters such as Low Probability of Intercept Radars, radios and laser designators. These are usually combined with active defenses such as chaff, flares, and ECM. It is accomplished by using a complex design philosophy to reduce the ability of an opponent's sensors to detect, track, or attack the stealth aircraft. This philosophy also takes into account the heat, sound, and other emissions of the aircraft as these can also be used to locate it.

Fig. 80. *A B-2 Spirit stealth bomber of the U.S Air Force.*

Fig. 81. A F-22 Raptor fifth generation stealth air superiority fighter.

Fig. 82. A prototype of the F-35 Lightning II fifth generation stealth multi-role fighter.

Full-size stealth combat aircraft demonstrators have been flown by the United States (in 1977), Russia (in 2010) and China (in 2011), while the US Military has already adopted three stealth designs, and is preparing to adopt another.

Most recent fighter designs will at least claim to have some sort of stealth, low observable, reduced RCS or radar jamming capability, but as of yet there has been no actual air to air combat experience against stealth aircraft.

During World War I, an attempt to reduce the visibility of military aircraft resulted in the German heavy bomber, the Linke-

Hofmann R.I; this had a wooden structure covered with transparent material.

The first true "stealth" aircraft may have been the Horten Ho 229 flying wing fighter-bomber, developed in Germany during the last years of World War II. In addition to the aircraft's shape, which may not have been a deliberate attempt to affect radar deflection, the majority of the Ho 229's wooden skin was bonded together using carbon-impregnated plywood resins designed with the purported intention of absorbing radar waves. Testing performed in early 2009 by the Northrop-Grumman Corporation established that this compound, along with the aircraft's shape, would have rendered the Ho 229 virtually invisible to Britain's Chain Home early warning radar, provided the aircraft was traveling at high speed (approximately 550 mph (890 km/h)) at extremely low altitude (50–100 feet).

In the closing weeks of WWII the US military initiated "Operation Paperclip", an effort by the US Army to capture as much advanced German weapons research as possible, and also to deny that research to advancing Soviet troops. A Horton glider and the Ho 229 number V3 were secured and sent to Northrop Aviation for evaluation in the United States, who much later used a flying wing design for the B-2 stealth bomber. During WWII Northrop had been commissioned to develop a large wing-only long-range bomber (XB-35) based on photographs of the Horton's record-setting glider from the 1930s, but their initial designs suffered controllability issues that were not resolved until after the war. Northrop's small one-man prototype (N9M-B) and a Horton wing-only glider are located in the Chino Air Museum in Southern California.

Modern stealth aircraft first became possible when Denys Overholser, a mathematician working for Lockheed Aircraft during the 1970s, adopted a mathematical model developed by Petr Ufimtsev, a Russian scientist, to develop a computer program called Echo 1. Echo made it possible to predict the radar signature an aircraft made with flat panels, called facets.

In 1975, engineers at Lockheed Skunk Works found that an aircraft made with faceted surfaces could have a very low radar signature because the surfaces would radiate almost all of the radar energy away from the receiver. Lockheed built a model called "the Hopeless Diamond", so-called because it resembled a squat diamond, and looked too hopeless to ever fly. Because advanced computers were available to control the flight of even a Hopeless Diamond, for the first time designers realized that it might be possible to make an aircraft that was virtually invisible to radar.

Reduced radar cross section is only one of five factors the designers addressed to create a truly stealthy design such as the F-22. The F-22 has also been designed to disguise its infrared emissions to make it harder to detect by infrared homing ("heat seeking") surface-to-air or air-to-air missiles. Designers also addressed making the aircraft less visible to the naked eye, controlling radio transmissions, and noise abatement.

The first combat use of purpose-designed stealth aircraft was in December 1989 during Operation Just Cause in Panama. On December 20, 1989, two USAF F-117s bombed a Panamanian Defense Force barracks in Rio Hato, Panama. In 1991, F-117s were tasked with attacking the most heavily fortified targets in Iraq in the opening phase of Operation Desert Storm and were the only jets allowed to operate inside Baghdad's city limits.

Early stealth aircraft were designed with a focus on minimal radar cross section (RCS) rather than aerodynamic performance. Highly-stealth aircraft like the F-117 Nighthawk are aerodynamically unstable in all three axes and require constant flight corrections from a fly-by-wire (FBW) flight system to maintain controlled flight. Most modern non-stealth fighter aircraft are unstable on one or two axes only. However, in the pursuit of increased maneuverability, most 4th and 5th-generation fighter aircraft have been designed with some degree of inherent instability that must be controlled by fly-by-wire computers. As for the B2 Spirit, based on the development of the all-wing aircraft by Jack Northrop since 1940, design allowed creating stable aircraft with sufficient yaw control, even without vertical surfaces such as rudders.

Earlier stealth aircraft (such as the F-117 and B-2) lack afterburners, because the hot exhaust would increase their infrared footprint, and breaking the sound barrier would produce an obvious sonic boom, as well as surface heating of the aircraft skin which also increased the infrared footprint. As a result their performance in air combat maneuvering required in a dogfight would never match that of a dedicated fighter aircraft. This was unimportant in the case of these two aircraft since both were designed to be bombers. More recent design techniques allow for stealthy designs such as the F-22 without compromising aerodynamic performance. Newer stealth aircraft, like the F-22 and F-35, have performance characteristics that meet or exceed those of current front-line jet fighters due to advances in other technologies such as flight control systems, engines, airframe construction and materials.

The high level of computerization and large amount of electronic equipment found inside stealth aircraft are often claimed to make them vulnerable to passive detection. This is highly unlikely and certainly systems such as Tamara and Kolchuga, which are often described as counter-stealth radars, are not designed to detect stray electromagnetic fields of this type. Such systems are designed to detect intentional, higher power emissions such as radar and communication signals. Stealth aircraft are deliberately operated to avoid or reduce such emissions.

Current Radar Warning Receivers look for the regular pings of energy from mechanically swept radars while fifth generation jet fighters use Low Probability of Intercept Radars with no regular repeat pattern.

Stealth aircraft are still vulnerable to detection during, and immediately after using their weaponry. Since stealth payload (reduced RCS bombs and cruise missiles) are not yet generally available, and ordnance mount points create a significant radar return, stealth aircraft carry all armament internally. As soon as weapons bay doors are opened, the plane's RCS will be multiplied and even older generation radar systems will be able to locate the stealth aircraft. While the aircraft will reacquire its stealth as soon as the bay doors are closed, a fast response defensive weapons system has a short opportunity to engage the aircraft.

This vulnerability is addressed by operating in a manner that reduces the risk and consequences of temporary acquisition. The B-2's operational altitude imposes a flight time for defensive weapons that makes it virtually impossible to engage the aircraft during its weapons deployment. All stealthy aircraft carry weapons in internal weapons bays. New stealth aircraft designs such as the F-22 and F-35 can open their bays, release munitions and return to stealthy flight in less than a second.

Some weapons require that the weapon's guidance system acquire the target while the weapon is still attached to the aircraft. This forces relatively extended operations with the bay doors open.

Also, such aircraft as the F-22 Raptor and F-35 Lighting II Joint Strike Fighter can also carry additional weapons and fuel on hardpoints below their wings. When operating in this mode the planes will not be nearly as stealthy, as the hardpoints and the weapons mounted on those hardpoints will show up on radar systems. This option therefore represents a trade off between stealth or range and payload. External stores allow those aircraft to attack more targets further away, but will not allow for stealth during that mission as compared to a shorter range mission flying on just internal fuel and using only the more limited space of the internal weapon bays for armaments.

Fully stealth aircraft carry all fuel and armament internally, which limits the payload. By way of comparison, the F-117 carries only two laser or GPS guided bombs, while a non-stealth attack aircraft can carry several times more. This requires the deployment of additional aircraft to engage targets that would normally require a single non-stealth attack aircraft. This apparent disadvantage however is offset by the reduction in fewer supporting aircraft that are required to provide air cover, air-defense suppression and electronic counter measures, making stealth aircraft "force multipliers".

The B-2 has a skin made with highly specialized materials such as Polygraphite.

Stealth aircraft are typically more expensive to develop and manufacture. An example is the B-2 Spirit that is many times more expensive to manufacture and support than conventional bomber aircraft. The B-2 program cost the U.S. Air Force almost $45 billion.

Theoretically there are a number of methods to detect stealth aircraft at long range.

Passive (multistatic) radar, bistatic radar and especially multistatic radar systems are believed to detect some stealth aircraft better than conventional monostatic radars, since first-generation stealth technology (such as the F117) reflects energy away from the transmitter's line of sight, effectively increasing the radar cross section (RCS) in other directions, which the passive radars monitor. Such a system typically uses either low frequency broadcast TV and FM radio signals (at which frequencies controlling the aircraft's signature is more difficult). Later stealth approaches do not rely on controlling the specular reflections of radar energy and so the geometrical benefits are unlikely to be significant.

Researchers at the University of Illinois at Urbana-Champaign with support of DARPA, have shown that it is possible to build a synthetic aperture radar image of an aircraft target using passive multistatic radar, possibly detailed enough to enable automatic target recognition (ATR).

In December 2007, SAAB researchers also revealed details for a system called Associative Aperture Synthesis Radar (AASR) that would employ a large array of inexpensive and redundant transmitters and a few intelligent receivers to exploit forward scatter to detect low observable targets. The system was originally designed to detect stealthy cruise missiles and should be just as effective against aircraft. The large array of inexpensive transmitters also provides a degree of protection against anti-radar (or anti-radiation) missiles or attacks.

Some analysts claim Infra-red search and track systems (IRSTs) can be deployed against stealth aircraft, because any aircraft surface heats up due to air friction and with a two channel IRST is a CO_2 (4.3 μm absorption maxima) detection possible, through difference comparing between the low and high channel. These analysts also point to the resurgence in such systems in several Russian designs in the 1980s, such as those fitted to the MiG-29 and Su-27. The latest version of the MiG-29, the MiG-35, is equipped with a new Optical Locator System that includes even more advanced IRST capabilities.

In air combat, the optronic suite allows:

Detection of non-afterburning targets at 45-kilometre (28 mi) range and more;

Identification of those targets at 8-to-10-kilometre (5.0 to 6.2 mi) range; and

Estimates of aerial target range at up to 15 kilometres (9.3 mi).

For ground targets, the suite allows:

A tank-effective detection range up to 15 kilometres (9.3 mi), and aircraft carrier detection at 60 to 80 kilometres (37 to 50 mi);

Identification of the tank type on the 8-to-10-kilometre (5.0 to 6.2 mi) range, and of an aircraft carrier at 40 to 60 kilometres (25 to 37 mi); and

Estimates of ground target range of up to 20 kilometres (12 mi).

The Dutch company Thales Nederland, formerly known as Holland Signaal, have developed a naval phased-array radar called SMART-L, which also is operated at L-Band and is claimed to offer counter stealth benefits. However, as with most claims of counter-stealth capability, these are unproven and untested. True resonant effects might be expected with HF sky wave radar systems, which have wavelengths of tens of metres. However, in this case, the accuracy of the radar systems is such that the detection is of limited value for engagement. Any radar which can successfully match the resonant frequency of a type of stealth aircraft should be able to detect its direction. In practice this is difficult because the resonant frequency changes depending on how the stealth aircraft is oriented with respect to the radar system.

Over-the-horizon radar is a design concept that increases radar's effective range over conventional radar. It is claimed that the Australian JORN Jindalee Operational Radar Network can overcome certain stealth characteristics. It is claimed that the HF frequency used and the method of bouncing radar from ionosphere overcomes the stealth characteristics of the F-117A. In other words, stealth aircraft are optimized for defeating much higher-frequency radar from front-on rather than low-frequency radars from above.

Stealth aircraft have been used in several conflicts: the United States invasion of Panama, the Gulf War, the Kosovo Conflict, the War in Afghanistan the War in Iraq and the 2011 military intervention in Libya. To date, the United States of America is the only country to have used stealth aircraft in combat.

The first use of stealth aircraft was in the United States invasion of Panama, where F-117 Nighthawk stealth attack aircraft were used to drop bombs on enemy airfields and positions while evading enemy radars.

The successful first deployment of stealth aircraft to a combat zone marks a milestone in military aviation.

In 1990 the F-117 Nighthawk was used again in the Gulf War, where F-117s flew approximately 1,300 sorties and scored direct hits on 1,600 high-value targets in Iraq while accumulating over 6,905 flight hours. Only 2.5% of the American aircraft in Iraq were F-117s, yet they struck more than 40% of the strategic targets, dropping over 2,000 tons of precision-guided munitions and striking their targets with over an 80% success rate.

In the 1999 NATO bombing of Yugoslavia two stealth aircraft were used by the United States, the veteran F-117 Nighthawk, and the newly introduced B-2 Spirit strategic stealth bomber.

The F-117 performed its usual role of striking precision high-value targets and performed well, although one F-117 was shot down by a Serbian Isayev S-125 'Neva-M' missile. The new B-2 Spirit was highly successful, destroying 33% of selected Serbian bombing targets in the first eight weeks of U.S. involvement in the War. During this war, B-2s flew non-stop to Kosovo from their home base in Missouri and back.

In the 2003 invasion of Iraq, F-117 Nighthawks and B-2 Spirits were again used, and this was the last time the F-117 would see combat.

F-117s dropped satellite-guided strike munitions on selected targets, with high success. B-2s conducted 49 sorties in the invasion, releasing more than 1.5 million pounds of munitions.

The most recent use of stealth aircraft was in the 2011 military intervention in Libya, where B-2 Spirits dropped 40 bombs on a Libyan airfield with concentrated air defenses in support of the UN no-fly zone.

In the future, it is likely that stealth aircraft will continue to play a valuable role in air combat. In future conflicts the United States is likely to use the F-22 Raptor, B-2 Spirit, and the F-35 Lightning II to perform a variety of operations.

In Russia, the Sukhoi PAK FA stealth multi-role fighter is scheduled to be introduced from 2015, to perform a wide variety of missions. In India, the Sukhoi/HAL FGFA, the Indian version of the PAK FA is scheduled to be introduced from 2017 in higher numbers, also to perform a wide variety of missions.

In the People's Republic of China, the Chengdu J-20 stealth multi-role fighter is planned to be introduced around 2018. A prototype was flown in early 2011.

The only time that a stealth aircraft has been shot down was on 27 March 1999, during Operation Allied Force. An American F-

117 Nighthawk's bomb bay had malfunctioned causing it to remain open for an unusually long time, allowing a Serbian Air Defense crew who were operating their radars on unusually long wavelengths to launch a Isayev S-125 'Neva-M' missile at it which brought it down. The pilot ejected and was rescued and the aircraft itself remained relatively intact due to striking the ground at a slow speed in an inverted position.

A B-2 crashed on 23 February 2008 shortly after takeoff from Andersen Air Force Base in Guam. The findings of the investigation stated that the B-2 crashed after "heavy, lashing rains" caused water to enter skin-flush air-data sensors, which feed angle of attack and yaw data to the computerized flight-control system.

The water distorted preflight readings in three of the plane's 24 sensors, causing the flight-control system to send an erroneous correction to the B-2 on takeoff. The B-2 quickly stalled, became unrecoverable, and crashed.

The sensors in question measure numerous environmental factors, including air pressure and density, for data to calculate airspeed, altitude and attitude. Because of the faulty readings, the flight computers determined inaccurate airspeed readings and incorrectly indicated a downward angle for the aircraft, which contributed to an early rotation and an un-commanded 30-degree pitch up and left yaw, resulting in the stall.

The Sukhoi PAK FA (fig. 83), (Perspektivny aviatsionny kompleks frontovoy aviatsii, literally "Prospective Airborne Complex of Frontline Aviation") is a twin-engine jet fighter being developed by Sukhoi OKB for the Russian Air Force.

The current prototype is Sukhoi's T-50 (fig. 84). The PAK FA, when fully developed, is intended to be the successor to the MiG-29 and Su-27 in the Russian inventory and serve as the basis of the Sukhoi/HAL FGFA being developed with India. A fifth generation jet fighter, the T-50 performed its first flight 29 January 2010. Its second flight was on 6 February and its third on 12 February 2010. As of 31 August 2010, it had made 17 flights and by mid-November, 40 in total. The second prototype was to start its flight test by the end of 2010, but this was delayed until March 2011.

Fig. 83. *A Sukhoi T-50 during a test flight.*

Fig. 84. *PAK FA T-50 prototype on the day of its first flight.*

Sukhoi director Mikhail Pogosyan has projected a market for 1,000 aircraft over the next four decades, which will be produced in a joint venture with India, 200 each for Russia and India and 600 for other countries. He has also said that the Indian contribution would be in the form of joint work under the current agreement rather than as a joint venture. The Indian Air Force will "acquire 50 single-seater fighters of the Russian version" before the two seat FGFA is developed. The Russian Defense Ministry will purchase the first 10 aircraft after 2012 and then 60 after 2016. The first batch of fighters will be delivered with current technology engines. Ruslan Pukhov, director of the Centre for Analysis of Strategies and Technologies, has projected that Vietnam will be the second export customer for the fighter. The PAK-FA is expected to have a service life of about 30–35 years.

In the late 1980s, the Soviet Union outlined a need for a next-generation aircraft to replace its MiG-29 and Su-27 in frontline service. Two projects were proposed to meet this need, the Sukhoi Su-47 and the Mikoyan Project 1.44. In 2002, Sukhoi was chosen to lead the design for the new combat aircraft.

The Tekhnokompleks Scientific and Production Center, Ramenskoye Instrument Building Design Bureau, the Tikhomirov Scientific Research Institute of Instrument Design, the Ural Optical and Mechanical Plant (Yekaterinburg), the Polet firm (Nizhniy Novgorod) and the Central Scientific Research Radio Engineering Institute (Moscow) were pronounced winners in the competition held in the beginning of 2003 for the development of the avionics suite for the fifth-generation airplane. NPO Saturn has been determined the lead executor for work on the engines for this airplane.

The Novosibirsk Chkalov Aviation Production Association (NAPO Chkalov) has begun construction of the fifth-generation multirole fighter. This work is being performed at Komsomol'sk-on-Amur together with Komsomolsk-on-Amur Aircraft Production Association; the enterprise's general director, Fedor Zhdanov reported during a visit to NAPO by Novosibirsk Oblast's governor Viktor Tolokonskiy on 6 March 2007. "Final assembly will take place at Komsomol'sk-on-Amur, and we will be carrying out assembly of the fore body of this airplane", Zhdanov specified.

On 8 August 2007, Russian Air Force Commander Alexander Zelin was quoted by Russian news agencies that the development stage of the PAK FA program is now complete and construction of the first aircraft for flight testing will now begin. Alexander Zelin also said that by 2009 there will be three fifth-generation aircraft ready. "All of them are currently undergoing tests and are more or less ready", he said. In the summer of 2009 the design was approved.

On 11 September 2010, it was reported that Indian and Russian negotiators had agreed on a preliminary design contract that would then be subject to Cabinet approval. The joint development deal would have each country invest $6 billion and take 8 to 10 years to develop the FGFA fighter. The agreement on the pre-design of the fighter was to be signed in December 2010. The preliminary design will cost $295 million and will be complete within 18 months.

On 28 February 2009, Mikhail Pogosyan announced that the airframe for the aircraft was almost finished and that the first prototype should be ready by August 2009. On 20 August 2009, Sukhoi General Director Mikhail Pogosyan said that the first flight would be by year end. Konstantin Makiyenko, deputy head of the Moscow-based Centre for Analysis of Strategies and Technologies said that "even with delays", the aircraft would likely make its first flight by January or February, adding that it would take 5 to 10 years for commercial production.

The maiden flight had been repeatedly postponed since early 2007 as the T-50 encountered unspecified technical problems. Air Force chief Alexander Zelin admitted as recently as August 2009 that problems with the engine and in technical research remained unsolved.

On 8 December 2009, Deputy Prime Minister Sergei Ivanov announced that the first trials with the fifth-generation aircraft would begin in 2010. The testing, however, has commenced earlier than stated, with the first successful taxiing test taking place on 24 December 2009.

The aircraft's maiden flight took place on 29 January 2010 at KnAAPO's Komsomolsk-on-Amur Dzemgi Airport; the aircraft was piloted by Sergey Bogdan (Сергей Богдан) and the flight lasted for 47 minutes.

A second airframe was first planned to join the flight testing in the fourth quarter of 2010 but was postponed. On 3 March 2011 a second prototype was reported to have made a successful 44 minutes test flight. These first two aircraft will lack radar and weapon control systems, while the third and fourth aircraft, to be added in 2011, will be fully functional test aircraft.

On 14 March 2011, the aircraft achieved supersonic flight at a test range near Komsomolsk-on-Amur in Siberia.

The T-50 is expected to be on display at the 2011 MAKS Airshow.

Navalized Sukhoi T-50 PAK FAs will be deployed on the Russian aircraft carrier Admiral Kuznetsov and future Russian aircraft carriers. There will be a competition between the Sukhoi, Mikoyan and Yakovlev design bureaus to choose the new naval aircraft.

Alexei Fedorov has said that any decision on applying fifth generation technologies to produce a smaller fighter (in the F-35 range) must wait until after the heavy fighter, based on the T-50, is completed.

Although most of information about the PAK FA is classified, it is believed from interviews with people in the Russian Air Force and Defense Ministry that it will be stealthy, have the ability to supercruise, be outfitted with the next generation of air-to-air, air-to-surface, and air-to-ship missiles, incorporate a fix-mounted AESA radar with a 1,500-element array and have an "artificial intellect".

According to Sukhoi, the new radar will reduce pilot load and the aircraft will have a new data link to share information between aircraft.

Composites are used extensively on the T-50 and comprise 25% of its weight and almost 70% of the outer surface. It is estimated that titanium alloy content of the fuselage is 75%. Sukhoi's concern for minimizing radar cross-section (RCS) and drag is also shown by the provision of two tandem main weapons bays in the centre fuselage, between the engine nacelles. Each is estimated to be between 4.9-5.1 m long. The main bays are augmented by bulged, triangular-section bays at the wing root.

The Moskovsky Komsomolets reported that the T-50 has been designed to be more maneuverable than the F-22 Raptor at the cost of making it less stealthy than the F-22. One of the design elements that have such an effect is the Leading Edge Vortex Controller (LEVCON).

The PAK FA SH121 radar complex includes three X-Band AESA radars located on the front and sides of the aircraft. These will be accompanied by L-Band radars on the wing leading edges. L-Band radars are proven to have increased effectiveness against very low observable (VLO) targets which are optimized only against X-Band frequencies, but their longer wavelengths reduce their resolution.

The PAK FA will feature an IRST optical/IR search and tracking system, based on the OLS-35M which is currently in service with the Su-35S.

Hindustan Aeronautics Limited will reportedly provide the navigation system and the mission computer.

The PAK FA was expected to use a pair of Saturn 117S engines on its first flights. The 117S (AL-41F1A) is a major upgrade of the AL-31F based on the AL-41F intended to power the Su-35BM, producing 142 kN (32,000 lb) of thrust in afterburner and 86.3 kN (19,400 lb) dry. In fact, PAK FA already used a completely new engine in its first flight, as stated by NPO Saturn. The engine is not based on the Saturn 117S and is rumoured to be called "127 engine". The engine generates a larger thrust and has a complex automation system, to facilitate flight modes such as maneuverability. Exact specifications of the new engine are still secret. It is expected that each engine will be able to independently vector its thrust upwards, downward or side to side. Vectoring one engine up with the other one down can produce a twisting force. Therefore the PAK FA would be the first fifth generation fighter with full 3-D thrust vectoring along all three aircraft axes: pitch, yaw and roll. These engines will incorporate infrared and RCS reduction measures.

The PAK FA has a reported maximum weapons load of 7,500 kg. It has an apparent provision for a cannon (most likely GSh-301). It could possibly carry as many as two 30 mm cannons. It has two internal bays estimated at 4.6-4.7 metres by 1-1.1 metres. Some sources suggest two auxiliary internal bays for short range AAMS and six external hardpoints.

Two Izdeliye 810 Extended beyond visual range missiles per weapons bay. Multiple Izdeliye 180 / K77M beyond visual range missiles. K74 and K30 within visual range missiles can also be carried. Two KH38M or KH58 USHK air-to-ground missiles per weapons bay. Multiple 250–500 kg precision guided bombs per weapons bay, with a maximum of 10 bombs in internal bays. Other possible loads include one 1,500 kg bomb per weapons bay or two 400 km+ range anti-AWACS weapons on external hardpoints.

The first flight video shows that PAK FA has no conventional rudders, its vertical tails are fully movable. This special tail fin design is mechanically similar to V-tails used by the Northrop YF-23 in 1990s, but is supplemented by dedicated horizontal stabilators (as on the F-22). The T-50 has wing leading-edge devices above the jet engine intakes that have been called a challenge for signature control.

The Chengdu J-20 (literally "Fighter aircraft Twenty", fig. 85) is a fifth generation stealth, twin-engine fighter aircraft prototype developed by Chengdu Aircraft Industry Group for the Chinese People's Liberation Army Air Force. In late 2010, the J-20 underwent high speed taxiing tests. The J-20 made its first flight on 11 January

2011. General He Weirong, Deputy Commander of the People's Liberation Army Air Force said in November 2009 that he expected the J-20 to be operational in 2017–2019.

Fig. 85. *A Chengdu J-20.*

The J-20 was one of the stealth fighter programs under the codename J-XX that was launched in the late 1990s. It was designated "Project 718", and won the PLAAF endorsement in a 2008 competition against a Shenyang proposal that was reportedly even larger than J-20. Two prototypes (2001-01 & 2001–02) have been built as of the end of 2010.

On 22 December 2010, the J-20 was under-going high speed taxiing tests outside the Chengdu Aircraft Design Institute with no confirmed flight tests. The J-20 made its first flight, which lasted about 20 minutes, on 11 January 2011.

Director of National Intelligence James R. Clapper has testified that the United States has known about the program for a "long time" and that the test flight was not a surprise.

The J-20 made its first flight, lasting about 15 minutes, on 11 January 2011. A Chengdu J-10S served as the escort aircraft. After the successful first flight, a ceremony was held. The test pilot of the J-20, Li Gang, Chief designer Yang Wei and General Li Andong (Deputy-Director of General Armaments Department, and Director of Science and Technology Commission of General Armaments Department of the PLA since 2000) attended the ceremony.

China thus became the third nation in the world to "develop and test-fly a full-size stealth combat aircraft demonstrator", after the United States and Russia. The Guardian reported that experts, on the one hand, expressed "surprise" at the speed with which the aircraft was developed, but on the other hand "said the country's military prowess was still relatively backward and way behind that of the US" and that its military interests were limited to its region.

The first test flight coincided with a visit of United States Secretary of Defense Robert Gates to China, and was initially interpreted by Pentagon officials and media pundits as a possible signal to the visiting delegation from the U.S. However, after meeting with senior Chinese officials including Chinese President Hu Jintao, Secretary Gates remarked, "The civilian leadership seemed surprised by the test and assured me it had nothing to do with my visit." Jin Canrong, a professor at Renmin University in Beijing who specializes in China-U.S. relations, suggested that President Hu's ignorance of the test raises questions about the nature of civilian control of the Chinese military. However, as Michael Swaine, an expert on the PLA and United States – China military relations, explained, although it's possible and even likely that "senior officials in the [Chinese] leadership did not know that this flight test would occur on this precise day," this is not necessarily evidence of a military-backed effort to insult Secretary Gates' delegation or embarrass President Hu. Rather, decisions regarding the production, development and testing of such military aircraft are routinely managed by engineers and low-level officials more than by senior civilian or military leadership. Coupled with the fact that there was relatively limited coverage of the event in Chinese media initially, it is likely that the test may not have been considered a significant enough event to warrant notification to President Hu. Moreover, the Chinese military has conducted important tests (including the 2007 anti-satellite missile test) on 11 January in the past; thus, the test may have been related to this.

A second test flight of an hour and twenty minutes took place on 17 April 2011. On 5 May 2011, a 55 minute test flight included retraction of the landing gear.

The full initial test program of 10 to 20 test flights is expected to take years to complete.

Globalsecurity.org states that China probably declined to participate in joint development and production of new fifth generation fighter with Russia given the belief that Russia stood to gain more from Chinese participation. Chinese leaders may have determined that their design was superior to the Russian PAK FA. United States House Committee on Armed Services chairman Howard McKeon said on the J-20 "my understanding is that they built it on information that they received from Russia, from a Russian plane, that they were able to copy".

Balkan military officials told the Associated Press that China and Russia may have adopted some stealth technology from a Lockheed F-117 Nighthawk, which was shot down by the Serbian military in 1999 during the Kosovo war. If Chinese experts used the F-117 stealth coatings, the result would be decades behind current American state-of-the-art. However, Chinese test pilot Xu Yongling said that the J-20 was a "masterpiece" of home-grown innovation, he also said the F-117 technology was already "outdated" even at the time it was shot down, and could not be applied to a next-generation stealth jet. Janes editor James Hardy agrees that it was unlikely China would have learned much from the wreckage.

Retired USAF General Thomas G. McInerney has suggested that the J-20 design may have been based on cyber-espionage of the Lockheed Martin FB-22 project.

A federal prosecutor has suggested that China may have used technology from the Northrop Grumman B-2 Spirit for their stealth aircraft which was supplied by Noshir Gowadia.

Chief of the Air Staff of the Indian Air Force Pradeep Vasant Naik has suggested that the J-20 is entirely reverse engineered with no Chinese R&D involved, and questioned if the practice was ethical. The Deccan Chronicle has called Naik's comment an "unusual outburst of helplessness" as China surpasses Indian airpower.

Russian military commentator Ilya Kramnik conjectures that China is still 10 to 15 years behind the United States and Russia in fighter technology and may not be able to manufacture all the advanced composite materials, avionics and sensor packages needed for such aircraft, and could instead turn to foreign suppliers. However, he speculates that China may be able to produce the J-20 at a cost 50% to 80% lower than US and Russian fifth-generation jet fighters, and that potential customers may include Pakistan, the Middle East, Latin America, Southeast Asia and the richest countries in Africa. Konstantin Sivkov of the Academy for Geopolitical Issues argued that the US is correct to be alarmed at the progress of Chinese military technology.

Bill Sweetman speculates that China will have problems meeting its production requirements, as it has several other jet fighter projects in production. Aviation Week raised the question of whether the aircraft is a prototype, like the Sukhoi T-50, or a technology demonstrator similar to the Lockheed YF-22.

The J-20 is a single-seat, twin-engine aircraft which appears to be somewhat larger and heavier than the comparable Sukhoi T-50 and Lockheed Martin F-22 Raptor. Bill Sweetman estimates that it is approximately 75 feet (23 m) in length, has a wingspan of 45 feet (14 m) or more, and is expected to have a takeoff weight of 75,000 to 80,000 pounds (34,000 to 36,000 kg) with internal stores only. The prototype could be powered by twin 32,000 pounds (15,000 kg) thrust Saturn 117S engines provided by Russia, a sign of problems in the development of the aircraft, according to Pentagon spokesman Col. David Lapan. Chinese sources have claimed that production aircraft will be powered by two 13,200 kilograms (29,000 lb)/WS-10 class high thrust turbofan engines fitted with Thrust Vector Controlled (TVC) nozzles, both made in China. However Richard Aboulafia has said that the WS-10 engine has suffered catastrophic failures in flight.

The J-20 may have lower supercruise speed (yet greater range) and less agility than a Lockheed Martin F-22 Raptor or PAK FA, but might also have larger weapons bays and carry more fuel. The J-20 has a long and wide fuselage and low jet engine intakes with a forward chine, a main delta wing, forward canards, a bubble canopy, conventional round engine exhausts and canted all-moving fins. The front section of the J-20 is similarly chiseled as the F-22 Raptor and the body and tail resemble those of the Sukhoi T-50 prototype. As early photographs of the prototype surfaced, Bill Sweetman commented that the design may suggest a large, long range ground attack aircraft, not unlike a "stealth version" of the General Dynamics F-111 Aardvark. Douglas Barrie has noted that the canard-delta configuration with canted vertical fins appears to resemble the MiG 1.42. Yet, Barrie notes that key differences include greater forward fuselage shaping as the basis for low observable characteristics, along with the different engine intake configuration. It is suspected that cyberespionage may have assisted the development of the J-20, with information used by subcontractors of Lockheed Martin for the F-35 project in particular having been significantly compromised during development of the J-20.

The J-20 has a pair of all-moving tailfins that are swept back in the F-35 style instead of being trapezoid like the F-22 and PAK-FA tails and ventral stabilizing fins. It also has an F-22 style nose section, but with F-35 style dropped nose, forward swept intake cowls with diverterless supersonic inlet (DSI) bumps and a one-piece canopy. It was reported in November 2006 that a T/W=10 17,000 kilograms (37,000 lb) class turbofan (WS-15/"large thrust") was being developed for the J-20. One (2001-01) prototype is fitted with AL-31F, the other (2001–02) is fitted with the improved WS-10G with a new "stealth" nozzle possibly to reduce RCS and IR emission.

The J-20 may become the first operational combat aircraft that carries sufficient fuel to supercruise throughout its missions, doubling its sortie rate.

Pentagon spokesman Geoff Morrell has said that it was premature to call the J-20 a stealth fighter or to judge if it had any other fifth generation characteristics.

The production J-20 may incorporate an advanced fly-by-wire (FBW) system fully integrated with the fire-control and the engine systems. Its fire-control radar is expected to be Active Electronically Scanned Array (AESA) (Type 1475/KLJ5?).

According to recent pictures from the internet, two small dark diamond shaped windows can be seen on both sides of the nose, which could house certain EO sensors, such as MAWS and/or IRST. Two additional windows are seen underneath the rear fuselage, plus two more on top of the forward fuselage above the canard wings, suggesting a distributed situational awareness system similar to the EODAS onboard American F-35 was installed providing a full 360° coverage.

The aircraft features a "pure" glass cockpit (two large color liquid crystal display (LCD) and several smaller ones and a wide-angle holographic head-up display (HUD)). Many of these subsystems have been tested onboard J-10Bs to speed up the development.

The J-20 has a large belly weapon bay for short/long-range air-to-air missiles (AAM) (PL-10, PL-12C/D & PL-21) and two smaller lateral weapon bays behind the air inlets for short-range AAMs (PL-10).

Carlo Kopp has suggested that the J-20's overall stealth shaping is "without doubt considerably better" than the F-35 and PAK FA, but he agrees with others, such as Shih Hiao-wei of Defense International monthly and Bill Sweetman of Aviation Week, that some parts on the J-20 will challenge its ability to remain stealthy from all directions: "The aft fuselage, tailbooms, fins/strakes and axisymmetric nozzles are not compatible with high stealth performance, but may only be stop-gap measures to expedite flight testing of a prototype." As of January 2011 the engine nozzles were clearly non-stealthy; this may be due to the fact that the final "fifth generation" engines had not been completed yet. However, one of the prototypes uses WS-10G engines with stealthy jagged-edge nozzles and tiles, however without the reduced RCS afterburners of the F119 and F135 this would have limited impact.

Robert Gates has also questioned how stealthy the J-20 might be although he did say the development of the J-20 had the potential to "put some of our capabilities at risk, and we have to pay attention to them, we have to respond appropriately with our own programs." Kopp and Goon have further speculated that the J-20 is designed to operate as a heavy interceptor, destroying opposing AWACS and tanker aircraft. If true, this would make it more similar to a MiG-25 with stealth capability. Sweetman agrees that this is the most likely role for such a large aircraft with low thrust to weight ratio and limited agility that is optimized for range and speed. Lewis Page has said that it is unlikely that the Chinese will soon have an American style Low Probability of Intercept Radar and so the J-20 would be limited to attacking ground targets like previous generations of American stealth aircraft such as the Lockheed F-117 Nighthawk. In that case the J-20 would carry a radar, but using it would instantly give away its location. However, the J-20 is expected to use a AESA radar, which should have Low Probability of Intercept modes. Given that the F-35 can already track and jam even the F-22's radar, this might not be sufficient.

Loren B. Thompson has said that this combination of forward sector only stealth and long range will allow the J-20 to make attacks on surface targets while the United States lacks sufficient bases for F-22s in the area to counter these attacks and American allies have no comparable aircraft. Thompson has also said that a long-range maritime strike aircraft may cause the United States more trouble than a shorter range air-superiority fighter like the F-22.

A canard delta offers greater efficiency in both subsonic and supersonic flight (which may help supercruise range), but it is unknown if the Chinese have the same software used on the Eurofighter Typhoon to control the otherwise non-stealthy canards. Teal Group analyst Richard Aboulafia has also raised doubts about the use of canards on a design that is intended to be low-observable: "There's no better way of guaranteeing a radar reflection and compromise of stealth". Aboulafia has also called the J-20 a kludge made of mismatched parts and questioned if the Chinese have the skills or technology to produce a true fifth generation fighter. Nevertheless, canards greatly boost the aircraft's maneuverability over that of a pure delta wing without canards. Sweetman notes that the canard delta works with the Whitcomb area rule for a large-volume mid-body section supersonic aircraft. Also, while the DSI intakes are easier to maintain than more complex stealth-compatible intakes, such as on the F-22, their fixed form limits the aircraft to around Mach 2.0. J.D. McFarlan of Lockheed Martin has said that the J-20 DSI inlets resemble those of the F-35, but it is unclear if the Chinese have perfected their own design.

The Advanced Medium Combat Aircraft (AMCA), formerly known as the Medium Combat Aircraft (MCA), is a single-seat, twin-engine fifth-generation stealth multirole fighter being developed by India. It will complement the HAL Tejas, the Sukhoi/HAL FGFA, the Sukhoi Su-30MKI and the as yet undecided MRCA in the Indian Air Force. The main purpose of this aircraft is to replace the aging SEPECAT Jaguar & Dassault Mirage 2000. Unofficial design work on the MCA has been started. A naval version is confirmed as Indian Navy also contributed to the funding. $2 billion funding is set to be allocated over the next three years.Number of AMCA orders are expected to reach 250 units.

In August 2006, India's then defence minister Mr. Pranab Mukherjee announced in Parliament that the government is evaluating experiences gained from the Tejas programme for the MCA.

In October 2008, the Indian Air Force asked the Aeronautical Development Agency (ADA) to prepare a detailed project report on the development of a Medium Combat Aircraft (MCA) incorporating stealth features.

In February 2009, ADA director P.S Subramanyam said at a Aero-India 2009 seminar, that they are working closely with Indian Air Force to develop a Medium Combat Aircraft. He added that according to the specification provided by the Indian Air Force, it would likely be a twenty ton aircraft powered by two GTX Kaveri engines.

In April 2010, the Indian Air Force issued the Air Staff requirements (ASR) for the AMCA which placed the aircraft in the twenty five ton category.

The AMCA will be designed with a very small radar cross-section and will also feature serpentine shaped air-intakes, internal weapons and the use of composites and other materials.

It will be a twin-engined design using the GTX Kaveri engine with thrust vectoring with the possibility of giving the aircraft supercruise capabilities. A wind-tunnel testing model of the MCA airframe was seen at Aero-India 2009.

As well as advanced sensors the aircraft will be equipped with missiles like DRDO Astra and other advanced missiles, stand-off weapons and precision weapons. The aircraft will have the capability to deploy JDAM's. The aircraft will feature Extended detection range and targeting range with the ability to release weapons at supersonic speeds. The aircraft's avionics suite will include AESA radar IRST and appropriate Electronic warfare systems and all aspect missile warning suite.

DARE, Bangalore has appointed a special team to begin identifying avionics and cockpit packages for the first prototype vehicle, and will supply this in published form to the ADA by July 2010. This will include cockpit electronics, cockpit configuration, man-machine interface, mission console systems and computers/software with a focus on data fusion and modular architecture. The LRDE will, in about the same time frame, provide a separate project proposal for an all new radar, to be re-designated for the AMCA, as a derivative of the MMR currently being completed with technology from Israel's ELTA. LRDE will independently look in the market for a partner for active array technology, though it communicated to ADA in June 2009 that it had sufficient R&D available to build a reliable AESA prototype with assistance from Bharat Electronics Ltd and two private firms based in Hyderabad.

The Next-Generation Bomber program (formerly called the 2018 Bomber) is a medium bomber under development by the United States Air Force. It was originally projected to enter service around 2018 as a super stealthy, subsonic, medium range, medium payload "B-3" type system to augment and possibly to a limited degree replace the U.S. Air Force's aging bomber fleet.

On 24 June 2010 Lt. Gen. Philip M. Breedlove said that the term "next-generation bomber" was dead and that the Air Force was working on a long-range strike "family" that would draw on the capabilities of systems like the F-35 and F-22 to help a more affordable and versatile bomber complete its missions.

On 13 September 2010 Air Force Secretary Michael Donley said that long range strike would continue cautiously with proven technologies and that the plan to be submitted with the 2012 budget could call for either a missile or an aircraft. General Norton Schwartz clarified that the bomber will not itself be nuclear capable, but will be the basis of a future nuclear capable aircraft.

USAF Air Combat Command in 2004-06 studied alternatives for a new bomber type aircraft to augment the current bomber fleet which now consists of largely 1970s era airframes, with a goal of having a fully operational aircraft on the ramp by 2018. Speculation that the next generation bomber would be hypersonic and unmanned were laid to rest when Air Force Major General Mark T. Matthews, head of ACC Plans and Programs said "Our belief is that the bomber should be manned" at a 1 May 2007 Air Force Association sponsored event. He later cited that the bomber would also likely be subsonic due to the higher cost of development and maintenance of a supersonic or hypersonic bomber. The 2018 bomber is expected to serve as a stop-gap until the more advanced "2037 Bomber" enters service.

USAF officials expect the new bomber to have top end low observability characteristics with the ability to loiter for hours over the battlefield area and respond to threats as they appear. Major General David E. Clary, ACC vice-commander, summed it up by saying the new bomber will be expected to "penetrate and persist". Deployment of cruise missiles is another issue for the new bomber. The B-52 is the only aircraft currently in the Air Force inventory allowed under treaty to carry and fire the cruise missiles. Major consideration was paid to operation readiness and flexibility. In 2006, the program expected that a prototype could be flying as early as 2009. In September 2007, Air Force generals stated that even though the development schedule for the bomber is short, it could be fielded by 2018.

On 25 January 2008, Boeing and Lockheed Martin announced an agreement to embark on a joint effort to develop a new U.S. Air

Force strategic bomber, with plans for the new airplane to be in service by 2018. This collaborative effort for a long-range strike program will include work in advanced sensors and future electronic warfare solutions, including advancements in network-enabled battle management, command and control, and virtual warfare simulation and experimentation. Under the Boeing-Lockheed arrangement, Boeing, the No. 2 Pentagon supplier, would be the primary contractor with about 60% of the deal, said sources familiar with the companies' plans. Lockheed, the world's largest defense contractor, would have around 40%. However on March 1, 2010 Boeing said that the joint project had been suspended. Northrop Grumman received $2 billion in funding in 2008 for "restricted programs" – also called black programs – for a demonstrator which could fly in 2010.

The Air Force was expected to announce late in 2009 its precise requirements for a new bomber that would be operating by 2018. In May 2009 testimony before Congress, U.S. Secretary of Defense Robert Gates mentioned that the Pentagon is considering a pilotless aircraft for the next-generation bomber role. Then in April 2009, Defense Secretary Gates announced a delay in the new generation bomber project that would push it past the 2018 date. This was caused not only by budget considerations, but also by nuclear arms treaty considerations.

On 19 May 2009, Air Force Chief of Staff General Norton Schwartz said that the USAF's focus in the 2010 budget was on "Long-range strike, not next-generation bomber" and will push for this in the QDR. In June 2009, the two teams working on NGB proposals were told to "close up shop".

On 16 September 2009, Defense Secretary Gates endorsed the concept of a new bomber but insisted that it must be affordable. He said, "I am committed to seeing that the United States has an airborne long-range strike capability – one of several areas being examined in the ongoing Quadrennial Defense Review. What we must not do is repeat what happened with our last manned bomber. By the time the research, development, and requirements processes ran their course, the aircraft, despite its great capability, turned out to be so expensive – $2 billion each in the case of the B-2 – that less than one-sixth of the planned fleet of 132 was ever built." On 5 October 2009, Ashton Carter said that the DoD was still deciding if the Air Force really needed a new bomber and that if the program was approved the aircraft would need to handle reconnaissance as well as strike. And in July 2010 he said he intended to "make affordability a requirement" for the next-generation intelligence and strike platform.

On 11 December 2009, Gates said that the QDR had shown the need for both manned and unmanned long range strike and that the 2011 budget would most likely include funding for the future bomber. The Air Force plans for the new bomber to be multi-role with intelligence, surveillance, and reconnaissance (ISR) capabilities.

Andrew Krepinevich has questioned the reliance on a short range aircraft like the F-35 to 'manage' China in a future conflict and has called on reducing the F-35 buy in favor of a longer range platform like the Next-Generation Bomber, but then-United States Secretary of the Air Force Michael Wynne rejected this plan of action back in 2007.

During the debate on the New START treaty in December 2010, several senators used the stalled bomber project as a reason to oppose or delay the ratification of the treaty.

US Secretary of Defense Robert Gates in a 6 January 2011 speech on the U.S. defense budget for FY 2012, announced that major investments will be made in developing a new, long-range, nuclear-capable penetrating bomber that has the option of being remotely piloted. He also said the aircraft "will be designed and developed using proven technologies, an approach that should make it possible to deliver this capability on schedule and in quantity. It is important that we begin this project now to ensure that a new bomber can be ready before the current aging fleet goes out of service.

The follow on bomber represents a key component of a joint portfolio of conventional deep-strike capabilities — an area that should be a high priority for future defense investment given the anti-access challenges our military faces". In addition to the strategic bombing, tactical bombing, and prompt global strike roles that one might expect for a long-range bomber, the new next generation bomber will be a part of a family of systems also responsible for ground surveillance and electronic attack.

The Air Force intends to purchase from 80 to 100 of the aircraft. Global Strike Command has indicated that one of the objectives for the bomber is for it to carry a weapon with the effects of the Massive Ordnance Penetrator. This would either be with the same weapon or a smaller weapon designed to match the penetrating power of the larger weapon.

The Obama administration in its 2012 budget request asked for $197 million and a total of $3.7 billion over five years to develop the bomber which would include modular payload options for Intelligence, Surveillance, Reconnaissance (ISR), Electronic Attack (EA), and communications. In 2011 the House Armed Services Committee added language that would require two engine programs for the bomber, but Ashton Carter objected that this would interfere with plans to reuse an existing engine.

In May 2011 Air Force undersecretary Erin Conaton announced that a program office for the bomber was being stood up.

The PAK DA (or PAK-DA), is a next generation strategic bomber which is being developed by Kazan Aircraft Production Association for Russia. It stands for Perspektivnyi Aviatsionnyi Kompleks Dalney Aviatsyi, which means Prospective Air Complex for Long Range Aviation. The PAK DA will be a new, stealthy, strategic bomber and is expected to enter service in the 2025–30 timeframe.

The Russian Air Force has tactical and technical requirements for a new generation of strategic bombers, as reported by Interfax. According to some sources, the PAK DA will be based on the supersonic Tu-160 bomber. Later references to the new bomber, including a televised address from Prime Minister Vladimir Putin, seem to imply the aircraft will be an entirely new design. Some speculation suggests that it might follow the stealthy design of the America B-2 Spirit bomber, but there is little public evidence to support that.

Russian Maj. Gen. Anatoly Zhikharev has stated that the new bomber will replace both the turboprop-powered Tupolev Tu-95 and the supersonic Tupolev Tu-160.

The 2037 Bomber is the unofficial name given to a heavy strategic bomber planned by the United States Air Force. It is projected to enter service in 2037 as a stealthy, supersonic, long-range, heavy-payload, possibly unmanned aircraft to potentially replace the B-52 Stratofortress, which is scheduled to be retired in 2040.

With the ending of B-2 production in the early 1990s, the U.S. Air Force was left with a gap in its bomber development. A new bomber would be needed in the 2037 time frame to replace retiring B-52s and B-1s according to the Air Force's Bomber Roadmap, released in 1999. This was considered too long to wait, as the Air Force needed an interim bomber before the 2030s. This left the aircraft proposed for 2037 as more of a heavy bomber than a medium bomber.

The McDonnell Douglas/General Dynamics A-12 Avenger II (fig. 86) was a proposed American ground-attack aircraft from McDonnell Douglas and General Dynamics. It was to be an all-weather, carrier-based stealth bomber replacement for the Grumman A-6 Intruder in the United States Navy and Marine Corps. Its Avenger II name was taken from the Grumman TBF Avenger of World War II.

The development of the A-12 was troubled by cost overruns and several delays, causing questions of the program's ability to deliver upon its objectives; these doubts led to the development program being canceled in 1991. The manner of its cancellation has been contested through litigation to this day.

Fig. 86. *A concept of the A-12 Avenger II.*

The US Navy began the Advanced Tactical Aircraft (ATA) program in 1983. The program was to develop and field a replacement for the A-6 Intruder by 1994. Stealth technology developed for the US Air Force would be used heavily in the program. Concept design contracts were awarded to the industry teams of McDonnell Douglas/General Dynamics, and Northrop/Grumman/Vought in November 1984. The teams were awarded contracts for further concept development in 1986.

The McDonnell Douglas/General Dynamics team was selected as the winner on 13 January 1988, the rival team lead by Grumman surprisingly failed to submit a final bid. The McDonnell Douglas/General Dynamics team was awarded a development contract and the ATA aircraft was designated A-12. The first flight was initially planned for December 1990. The A-12 was named Avenger II in homage to the World War II-era Navy torpedo-bomber Grumman TBF Avenger.

The Navy initially sought to buy 620 A-12s and Marines wanted 238. In addition, the Air Force briefly considered ordering some 400 of an A-12 derivative. The A-12 was promoted as as possible replacement for the Air Force's General Dynamics F-111 Aardvark, and for the United Kingdom's Panavia Tornado fighter-bombers. The craft was a flying wing design in the shape of an isosceles triangle, with the cockpit situated near the apex of the triangle. The A-12 gained the nickname "Flying Dorito".

The aircraft was to be powered by two General Electric F412-D5F2 turbofan engines, each producing about 13,000 pounds-force (58 kN) of thrust. It was designed to carry precision guided weapons internally, up to two AIM-120 AMRAAM air-to-air missiles, two AGM-88 HARM air-to-ground missiles and a complement of air-to-ground ordnance, including unguided or precision-guided bombs, could be carried in an internal weapons bay. It has been claimed that the A-12 was to be capable of delivering nuclear weapons held in its internal weapons bay as well. The A-12 was to have a weapons load of 5,160 pounds (2,300 kg).

Beginning in early 1990 McDonnell Douglas and General Dynamics revealed delays and projected cost increases. The weight of the aircraft had significantly increased due to complications with the composite materials used, the weight being 30% over design specification, this was a significantly negative factor for carrier-based operations. Technical difficulties with the complexity of the radar system to be used also caused costs to increase; by one estimate the A-12 was to consume up to 70% of the Navy's budget for aircraft. After delays, its critical design review was successfully completed in October 1990; the A-12's maiden flight was rescheduled to early 1992. In December 1990, it was planned for 14 Navy aircraft carriers to equipped with a wing of 20 A-12s each.

A government report released in November 1990 documented serious problems with the A-12 development program. In December 1990 Secretary of Defense Dick Cheney told the Navy to justify the program and deliver reasons why it should not be canceled. The response given by the Navy and the contractors failed to persuade the Secretary of Defense, as he canceled the program in the following month, on 7 January 1991, for breach of contract.

The government felt the contractors could not complete the program and instructed them to repay most of the $2 billion that had been spent on A-12 development. McDonnell Douglas and General Dynamics disputed this in Federal Claims court; the reasons and causes for the cancellation have been debated and remain an issue of controversy, with suggestions of political expediency and scheming mooted to be behind the action.

After the cancellation of the A-12, the Navy elected to purchase the F/A-18E/F Super Hornet, which went on to replace the A-6 Intruder and the F-14 Tomcat. The Super Hornet uses the General Electric F414 turbofan engine, which is a modified variant of the upgraded F404 version developed for the A-12. The full-size A-12 mockup was revealed to the public at the former Carswell Air Force Base in June 1996. The cancellation of the A-12 is seen as one of the major losses in the 1990s that weakened McDonnell Douglas and led to its merger with rival Boeing in 1997.

The manner in which the program was canceled led to years of litigation between the contractors and the Department of Defense over breach of contract. On 1 June 2009, the U.S. Court of Appeals for the Federal Circuit ruled that the U.S. Navy was justified in canceling the contract. The ruling also required the two contractors to repay the U.S. government more than US$1.35 billion, plus interest charges of US$1.45 billion. Boeing, which had merged with McDonnell Douglas, vowed to appeal the decision, as has General Dynamics. In September 2010, the U.S. Supreme Court said it would hear the two companies' arguments, that the government canceled the project improperly and that the use of a state secrets claim by the U.S. prevented them from mounting an effective defense. In May 2011, the Supreme Court set aside the Appeals Court decision and returned the case to federal circuit court.

The Lockheed Martin F-35 Lightning II, an in-development carrier-capable fighter featuring stealth technology and oriented toward ground-attack operations, is in effect a successor to the A-12 in both role and industrial origins. Comparisons have been drawn between the role of the future Next-Generation Bomber and the A-12; the two aircraft both being stealthy sub-sonic bombers designed to carry comparative bomb loads and fly similar ranges.

The Mitsubishi ATD-X Shinshin is a state of the art prototype fifth-generation jet fighter that uses advanced stealth technology. Being developed by the Japanese Ministry of Defense Technical Research and Development Institute (TRDI) for research purposes. The main contractor of the project is Mitsubishi Heavy Industries. Many consider this aircraft to be Japan's first domestically made stealth fighter. ATD-X is an acronym meaning "Advanced Technology Demonstrator – X". The aircraft's Japanese name is 心神 (shin-shin) which means one's mind. The aircraft's first flight is scheduled for 2014.

At the beginning of the 21st century, Japan sought to replace its aging fleet of fighter aircraft, began making overtures to the United States on the topic of purchasing several Lockheed Martin F-22 Raptor fighters for their own forces. However the U.S. Congress had banned the exporting of the aircraft in order to safeguard secrets of the aircraft's technology such as its extensive use of stealth; this rejection necessitated Japan to develop its own modern fighter, to be equipped with stealth features and other advanced systems.

A mock-up of the ATD-X was constructed and used to study the radar cross section in France in 2005. A radio-controlled 1/5 scale model made its first flight in 2006 to gain data on performance at high angles of attack and to test new sensory equipment and self-repairing flight control systems.

Following these preliminary steps, the decision was taken in 2007 to push ahead with the multi billion-yen project. At the time of this decision, production was forecast to start roughly 10 years later, around 2017. As of 2007, the ATD-X is expected to conduct its maiden flight in 2014.

The ATD-X will be used as a technology demonstrator and research prototype to determine whether domestic advanced technologies for a fifth generation fighter aircraft are viable, and is a 1/3 size model of a possible full-production aircraft. The aircraft also features 3D thrust vectoring capability. Thrust is controlled in the ATD-X by the use of 3 paddles on each engine nozzle similar to the

186

system used on the Rockwell X-31, while an axis-symmetric thrust vectoring engine is also being developed for the full scale production model. The nozzles on the prototype appear to be uncovered and might have a slight adverse effect on the aircraft's stealth characteristics.

Among the features the ATD-X is to have is a fly-by-optics flight control system, which by substituting optical fibers for wires, allows data to be transferred faster and with immunity to electromagnetic disturbance.

Its radar will be an active electronically scanned array (AESA) radar called the 'Multifunction RF Sensor', which is intended to have broad spectrum agility, capabilities for electronic countermeasures (ECM), electronic support measures (ESM), communications functions, and possibly even microwave weapon functions.

A further feature will be a so-called 'Self Repairing Flight Control Capability', which will allow the aircraft to automatically detect failures or damage in its flight control surfaces, and using the remaining control surfaces, calibrate accordingly to retain controlled flight.

The JASDF is reported to have issued a request for information for engines in the 10 to 20 thousand pound thrust range to power the prototypes while Ishikawajima-Harima Heavy Industries is to provide the engines for the completed fighter.

The Lockheed Have Blue was the code name for Lockheed's "proof of concept" (i.e., prototype) stealth aircraft that preceded the F-117 Nighthawk production stealth aircraft program. Have Blue was designed by Lockheed's Skunk Works division, and tested at the top-secret Groom Lake base, Nevada. The Have Blue was the first fixed-wing aircraft designed from an electrical engineering (rather than an aerospace engineering) perspective. The aircraft's plate-like, faceted shape was designed to deflect electromagnetic waves, greatly reducing its radar signature.

Two Have Blue planes were built to test both the flight dynamics and radar returns of the stealth concept. These prototypes flew at Groom Lake, Nevada, between 1977 and 1979. While they appear similar to the later F-117, the Have Blue prototypes were smaller aircraft, about 60% scale, with greater wing sweep and inward-canted vertical tails. The nose of Have Blue prototypes was also sharper and offered a slightly higher degree of stealth compared to production F-117 planes, which had to have a flat windshield to incorporate a head-up display.

During testing of the design, the aircraft was flown near (~100 miles away) to an army radar system, followed at some significant distance by a spotter plane; over a preplanned flight path. The cover story for the technology was that a black box in the nose of the aircraft was able to deflect the radar; whereas obviously the shape of the aircraft did all the real work. Radar only managed to detect the spotter plane; a soldier placed on the ground directly under the flight path had to witness the oddly shaped plane to verify that the flight had occurred.

The design was inherently unstable about all three axes, control being fly-by-wire adapted from the F-16's single-axis fly-by-wire system. Both aircraft were ultimately lost in the course of testing, the first from a hard landing incident which resulted in the gear being jammed in a semi-retracted position and the pilot ultimately being ordered to eject after attempts to enable the gear to lower and lock proved unsuccessful. The second was the result of an engine fire which severed hydraulic lines, forcing the pilot to eject. The debris from both aircraft were secretly buried somewhere within the Nellis complex.

"Even though the test site was in a remote location, our airplane was kept under wraps inside its hangar most of the time. Soviet spy satellites made regular passes, and every time our airplane was rolled out everyone on the base who wasn't cleared for Have Blue had to go into the windowless mess hall and have a cup of coffee until we took off."

— Ben Rich, director of Lockheed's Skunk Works from 1975 to 1991.

The Northrop Tacit Blue was a technology demonstrator aircraft created to demonstrate that a stealth low observable surveillance aircraft with a low probability of intercept radar and other sensors could operate close to the forward line of battle with a high degree of survivability (fig. 87).

Unveiled by the U.S. Air Force on 30 April 1996, the Tacit Blue Technology Demonstration Program was designed to prove that such an aircraft could continuously monitor the ground situation deep behind the battlefield and provide targeting information in real-time to a ground command center. Tacit Blue represented the 'black' component in the larger Assault Breaker program, which intended to validate the concept of massed standoff attacks on advancing armoured formations using smart munitions. The Pave Mover radar demonstrators provided the non-stealthy portion of the program's targeting system, whereas Tacit Blue was intended to demonstrate a similar but stealthy capability, while validating a number of innovative stealth technology advances.

Fig. 87. The Northrop Tacit Blue was a Stealth demonstrator.

Tacit Blue, nicknamed "the whale," featured a straight tapered wing with a V-tail mounted on an oversized fuselage with a curved shape. A single flush inlet on the top of the fuselage provided air to two high-bypass turbofan engines. Tacit Blue employed a quadruply redundant, digital, fly-by-wire flight control system to help stabilize the aircraft about its longitudinal and directional axes.

The sensor technology developed for Tacit Blue is now being used by the E-8 Joint STARS aircraft.

The aircraft made its first flight in February 1982, and subsequently logged 135 flights over a three year period. The aircraft often flew three to four flights weekly and several times flew more than once a day. After reaching about 250 flight hours, the aircraft was placed in storage in 1985. In 1996, Tacit Blue was placed on display at the National Museum of the United States Air Force at Wright Patterson Air Force Base, near Dayton, Ohio. Tacit Blue is on display in the Research and Development Hangar (within the Wright-Patterson Air Force Base perimeter and away from the main National Museum site).

The Avro Vulcan, sometimes referred to as the Hawker Siddeley Vulcan, is a delta wing subsonic jet strategic bomber that was operated by the Royal Air Force (RAF) from 1953 until 1984. It was developed by Avro in response to a specification released by the Air Ministry. At the time, both jet engines and delta wings were considered cutting-edge and relatively unexplored; thus, the small-scale Avro 707 was produced to test the principles of the design. In flight, the Vulcan was an agile aircraft for its size.

The Vulcan B.1 was first delivered to the RAF in 1956. In service, the Vulcan was armed with nuclear weapons and was a part of the RAF's V bomber force, the United Kingdom's airborne deterrent against aggression from other powers such as the Soviet Union during the Cold War. Features such as an extensive electronic countermeasures suite and a low radar cross section for its size would have made the aircraft difficult to detect while carrying out the nuclear strike mission. A second batch of aircraft, the B.2, was produced with new features, including a larger wing and greater fuel capacity, along with more advanced electronics and radar systems.

The B.2s were adapted into several other variants, the B.2A carrying the Blue Steel missile, the B.2 (MRR) for Marine Radar Reconnaissance use, and the K.2 tanker for aerial refuelling. The Vulcan was also used in the secondary role of conventional bombing near the end of its service life in the 1982 Falklands War against Argentina during Operation Black Buck. One example, XH558, was recently restored for use in display flights and commemoration of the employment of the aircraft in the Falklands conflict.

Design work began at A. V. Roe in 1947 under Roy Chadwick, however, the delta wing design built upon the wartime work of Professor Alexander Lippisch, and the first design studies featured a radical tailless delta wing design. The Air Ministry specification B.35/46 required a bomber with a top speed of 575 mph (925 km/h), an operating ceiling of 50,000 ft (15,000 m), a range of 3,452 miles (5,556 km) and a bomb load of 10,000 lb (4,500 kg); intended to carry out delivery of Britain's nuclear-armed gravity bombs to strategic targets within Soviet territory. Design work also began at Vickers and Handley Page. All three designs were approved – aircraft that would become the Valiant, the Victor, and the Avro Vulcan.

The Type 698 as first envisaged by Chadwick, and upon his death in the crash of the Avro Tudor 2 prototype on 23 August 1947, later refined by his successor, Stuart Davies in March 1949, was a more conventional delta wing design initially with tail surfaces at the ends of the wing and finally with a "full" tail unit. Avro felt this would be able to give the required combination of large wing area and sweepback to offset the transonic effects and a thick wing root to embed the engines. The thick wing gave considerable space for the engines and made allowances for future larger models to be installed. Wingtip rudders provided the control instead of the traditional rear fuselage and tail, which were unnecessary on this design. This design was reworked multiple times to reduce weight, a restriction which was later loosened, and became more conventional, adopting a centre fuselage with four paired engines and a tail.

As the delta wing was an unknown quantity, Avro began scale prototype testing in 1948 with a series of single-seater Avro 707 "proof-of-concept" aircraft. Despite the crash of the first 707 prototype on 30 September 1949, work continued and eventually four additional 707s were built to test low-speed as well as high-speed characteristics of the delta wing design.

The first full-scale prototype Type 698 made its maiden flight piloted by Avro Chief Test Pilot retired Wing Commander Roly Falk on 30 August 1952. The Vulcan name was not chosen until 1952, after the Valiant had already been named. The 698's fourth flight was its appearance at the 1952 Farnborough airshow when Roly Falk demonstrated an "almost vertical bank". Falk made another public demonstration of the Vulcan's high performance and agility at the 1955 Farnborough Airshow; while flying the second production Vulcan, XA890, he performed an upward barrel roll shortly following takeoff. After the second occasion he rolled it at the show, the SBAC requested he stop as it was "inappropriate".

Two prototypes were built and subsequently modified for development purposes. Both flew with a straight leading edge, which was then modified to have a kink further out towards the wingtip; this was to remove the occurrence of high frequency buffeting during flight. The production Vulcan bomber in service was fitted with a delta wing, but these were not pure delta wings for reason of better flying characteristics, and successive variants saw the wing altered and expanded again.

Like the rest of the V-bombers no foreign country purchased Vulcans, however interest had been present. As early as 1954, Australia recognised that the English Electric Canberra was becoming outdated and evaluated aircraft such as the Avro Vulcan and Handley-Page Victor as potential replacements. However political pressure for a Canberra replacement only rose to a head in 1962; at which point more modern types such as the BAC TSR-2, General Dynamics F-111C, and North American A-5 Vigilante had become available, ultimately the F-111C was procured.

The RQ-5 Hunter unmanned aerial vehicle (UAV) was originally intended to serve as the United States Army's Short Range UAV system for division and corps commanders. It took off and landed (using arresting gear) on runways. It used a gimbaled EO/IR sensor to relay its video in real time via a second airborne Hunter over a C-band line-of-sight data link. The RQ-5 is based on the Hunter UAV that was developed by Israel Aircraft Industries.

General Dynamics F-16 Fighting Falcon

The General Dynamics F-16 Fighting Falcon is a multirole jet fighter aircraft originally developed by General Dynamics for the United States Air Force (USAF). Designed as a lightweight day fighter, it evolved into a successful all-weather multirole aircraft. Over 4,400 aircraft have been built since production was approved in 1976. Though no longer being purchased by the U.S. Air Force, improved versions are still being built for export customers.

In 1993, General Dynamics sold its aircraft manufacturing business to the Lockheed Corporation, which in turn became part of Lockheed Martin after a 1995 merger with Martin Marietta.

The Fighting Falcon (fig. 88) is a dogfighter with numerous innovations including a frameless bubble canopy for better visibility, side-mounted control stick to ease control while maneuvering, a seat reclined 30 degrees to reduce the effect of g-forces on the pilot, and the first use of a relaxed static stability/fly-by-wire flight control system that makes it a highly nimble aircraft. The F-16 has an internal M61 Vulcan cannon and has 11 hardpoints for mounting weapons, and other mission equipment. Although the F-16's official name is "Fighting Falcon", it is known to its pilots as the "Viper", due to it resembling a viper snake and after the Battlestar Galactica Colonial Viper starfighter.

Fig. 88. *F-16 Fighting Falcon.*

In addition to USAF active, reserve, and air national guard units, the aircraft is used by the USAF aerial demonstration team, the U.S. Air Force Thunderbirds, and as an adversary/aggressor aircraft by the United States Navy. The F-16 has also been procured to serve in the air forces of 25 other nations.

Experience in the Vietnam War revealed the need for air superiority fighters and better air-to-air training for fighter pilots. Based on his experiences in the Korean War and as a fighter tactics instructor in the early 1960s Colonel John Boyd with mathematician Thomas Christie developed the Energy-Maneuverability theory to model a fighter aircraft's performance in combat. Boyd's work called for a small, lightweight aircraft with an increased thrust-to-weight ratio. In the late 1960s, Boyd gathered a group of like-minded innovators that became known as the Fighter Mafia and in 1969 they secured DoD funding for General Dynamics and Northrop to study design concepts based on the theory.

Air Force F-X proponents remained hostile to the concept because they perceived it as a threat to the F-15 program. However, the Advanced Day Fighter concept, renamed F-XX gained civilian political support under the reform-minded Deputy Secretary of Defense David Packard, who favored the idea of competitive prototyping. As a result in May 1971, the Air Force Prototype Study Group was established, with Boyd a key member, and two of its six proposals would be funded, one being the Lightweight Fighter (LWF). The Request for Proposals issued on 6 January 1972 called for a 20,000-pound (9,100 kg) class air-to-air day fighter with a good turn rate, acceleration and range, and optimized for combat at speeds of Mach 0.6–1.6 and altitudes of 30,000–40,000 feet (9,100–12,000 m). This was the region where USAF studies predicted most future air combat would occur. The anticipated average flyaway cost of a production version was $3 million. This production plan, though, was only notional as the USAF had no firm plans to procure the winner.

Five companies responded and in 1972, the Air Staff selected General Dynamics' Model 401 and Northrop's P-600 for the follow-on prototype development and testing phase. GD and Northrop were awarded contracts worth $37.9 million and $39.8 million to produce the YF-16 and YF-17, respectively, with first flights of both prototypes planned for early 1974. To overcome resistance in the Air Force hierarchy, the Fighter Mafia and other LWF proponents successfully advocated the idea of complementary fighters in a high-cost/low-cost force mix. The "high/low mix" would allow the USAF to be able to afford sufficient fighters for its overall fighter force structure requirements. The mix gained broad acceptance by the time of the flyoff between the prototypes, and would define the relationship of the LWF and the F-15.

The first YF-16 was rolled out on 13 December 1973, and its 90-minute maiden flight was made at the Air Force Flight Test Center (AFFTC) at Edwards AFB, California, on 2 February 1974. Its actual first flight occurred accidentally during a high-speed taxi test on 20 January 1974. While gathering speed, a roll-control oscillation caused a fin of the port-side wingtip-mounted missile and then the starboard stabilator to scrape the ground, and the aircraft then began to veer off the runway. The GD test pilot, Phil Oestricher, decided to lift off to avoid crashing the machine, and safely landed it six minutes later. The slight damage was quickly repaired and the official first flight occurred on time. The YF-16's first supersonic flight was accomplished on 5 February 1974, and the second YF-16 prototype first flew on 9 May 1974. This was followed by the first flights of the Northrop's YF-17 prototypes on 9 June and 21 August 1974, respectively. During the flyoff, the YF-16s completed 330 sorties for a total of 417 flight hours; the YF-17s flew 288 sorties, covering 345 hours.

Increased interest would turn the LWF into a serious acquisition program. North Atlantic Treaty Organization (NATO) allies Belgium, Denmark, the Netherlands, and Norway were seeking to replace their F-104G fighter-bombers. In early 1974, they reached an agreement with the U.S. that if the USAF ordered the LWF winner, they would consider ordering it as well. The USAF also needed to replace its F-105 and F-4 fighter-bombers. The U.S. Congress sought greater commonality in fighter procurements by the Air Force and Navy, and in August 1974 redirected Navy funds to a new Navy Air Combat Fighter (NACF) program that would be a navalized fighter-bomber variant of the LWF. The four NATO allies had formed the "Multinational Fighter Program Group" (MFPG) and pressed for a U.S. decision by December 1974. The U.S. Air Force then advanced its plans to announce the LWF winner from May 1975 to the beginning of the year, and accelerated testing.

To reflect this more serious intent to procure a new fighter-bomber design, the LWF program was rolled into a new Air Combat Fighter (ACF) competition in an announcement by U.S. Secretary of Defense James R. Schlesinger in April 1974. Schlesinger also made it clear that any ACF order would be for aircraft in addition to the F-15, which extinguished opposition to the LWF. ACF also raised the stakes for GD and Northrop because it brought in further competitors intent on securing the lucrative order that was touted at the time as "the arms deal of the century". These were Dassault-Breguet's Mirage F1M-53, the SEPECAT Jaguar, and a proposed derivative of the Saab 37 Viggen named the "Saab 37E Eurofighter". Northrop offered the P-530 Cobra, which was very similar to its YF-17. The Jaguar and Cobra were dropped by the MFPG early on, leaving two European and the two U.S. candidates. On 11 September 1974, the U.S. Air Force confirmed firm plans to place an order for the winning ACF design sufficient to equip five tactical fighter wings. On 13 January 1975, Secretary of the Air Force John L. McLucas announced that the YF-16 had been selected as the winner of the ACF competition.

The chief reasons given by the Secretary for the decision were the YF-16's lower operating costs, greater range and maneuver performance that was "significantly better" than that of the YF-17, especially at near-supersonic and supersonic speeds. Another advantage was the fact that the YF-16 – unlike the YF-17 – employed the Pratt & Whitney F100 turbofan engine, which was the same powerplant used by the F-15; such commonality would lower the unit costs of engines for both programs.

Shortly after selection of the YF-16, Secretary McLucas revealed that the USAF planned to order at least 650 and up to 1,400 of the production F-16 version. In the Navy Air Combat Fighter (NACF) competition, the Navy announced on 2 May 1975 that it selected the YF-17 as the basis for what would become the McDonnell Douglas F/A-18 Hornet.

The U.S. Air Force initially ordered 15 "Full-Scale Development" (FSD) aircraft (11 single-seat and four two-seat models) for its flight test program, but this was reduced to eight (six F-16A single-seaters and two F-16B two-seaters). The YF-16 design was altered for the production F-16. The fuselage was lengthened by 10.6 in (0.269 m), a larger nose radome was fitted to house the AN/APG-66 radar, wing area was increased from 280 sq ft (26 m2) to 300 sq ft (28 m2), the tailfin height was decreased slightly, the ventral fins were enlarged, two more stores stations were added, and a single side-hinged nosewheel door replaced the original double doors. These modifications increased the F-16's weight approximately 25% over that of the YF-16 prototypes.

Manufacture of the FSD F-16s got underway at General Dynamics' Fort Worth, Texas plant in late 1975, with the first example, an F-16A, being rolled out on 20 October 1976, followed by its first flight on 8 December. The initial two-seat model achieved its first flight on 8 August 1977. The initial production-standard F-16A flew for the first time on 7 August 1978 and its delivery was accepted by the USAF on 6 January 1979. The F-16 was given its formal nickname of "Fighting Falcon" on 21 July 1980, entering USAF operational service with the 388th Tactical Fighter Wing at Hill AFB on 1 October 1980.

On 7 June 1975, the four European partners, now known as the European Participation Group, signed up for 348 aircraft at the Paris Air Show. This was split among the European Participation Air Forces (EPAF) as 116 for Belgium, 58 for Denmark, 102 for the Netherlands, and 72 for Norway. These would be produced on two European production lines, one in the Netherlands at Fokker's Schiphol-Oost facility and the other at SABCA's Gossellies plant in Belgium; production would be divided among them as 184 and 164 units, respectively. Norway's Kongsberg Vaapenfabrikk and Denmark's Terma A/S also manufactured parts and subassemblies for the EPAF aircraft. European co-production was officially launched on 1 July 1977 at the Fokker factory. Beginning in mid-

November 1977, Fokker-produced components were shipped to Fort Worth for assembly of fuselages, which were in turn shipped back to Europe (initially to Gossellies starting in January 1978); final assembly of EPAF-bound aircraft began at the Belgian plant on 15 February 1978, with deliveries to the Belgian Air Force beginning in January 1979. The Dutch line started up in April 1978 and delivered its first aircraft to the Royal Netherlands Air Force in June 1979. In 1980 the first aircraft were delivered to the Royal Norwegian Air Force by SABCA and to the Royal Danish Air Force by Fokker.

Since then, a further production line has been established at Ankara, Turkey, where Turkish Aerospace Industries (TAI) has produced 232 Block 30/40/50 F-16s under license for the Turkish Air Force during the late 1980s and 1990s, and has 30 Block 50 Advanced underway for delivery from 2010; TAI also built 46 Block 40s for Egypt in the mid-1990s. Korean Aerospace Industries opened another production line for the KF-16 program, producing 140 Block 52s from the mid-1990s to mid-2000s. If India selects the F-16IN for its Medium Multi-Role Combat Aircraft procurement, a sixth F-16 production line will be established in that nation to produce at least 108 fighters.

One change made during production was the need for more pitch control to avoid deep stall conditions at high angles of attack, this issue was known about in development but had originally been discounted. Model tests of the YF-16 conducted by the Langley Research Center revealed a potential problem, but no other laboratory was able to duplicate it. YF-16 flight tests were not sufficient to expose the issue, it required later flight testing on the FSD aircraft to demonstrate there was a real concern. In response, the areas of the horizontal stabilizer were increased 25%; this so-called "big tail" was introduced on the Block 15 aircraft in 1981 and retrofitted later on earlier production aircraft. Besides significantly reducing (though not eliminating) the risk of deep stalls, the larger horizontal tails also improved stability and permitted faster takeoff rotation.

In the 1980s, the Multinational Staged Improvement Program (MSIP) was conducted to evolve new capabilities for the F-16, mitigate risks during technology development, and ensure the aircraft's worth. The program upgraded the F-16 in three stages. The MSIP process permitted the introduction of new capabilities quicker, at lower costs and with reduced risks, compared to traditional independent programs to upgrade and modernize aircraft. The F-16 has been involved in other upgrade programs including service life extension programs in the 2000s.

The F-16 is a single-engined, supersonic, multi-role tactical aircraft. The F-16 was designed to be a cost-effective combat "workhorse" that can perform various kinds of missions and maintain around-the-clock readiness. It is much smaller and lighter than its predecessors, but uses advanced aerodynamics and avionics, including the first use of a relaxed static stability/fly-by-wire (RSS/FBW) flight control system, to achieve enhanced maneuver performance. Highly nimble, the F-16 can pull 9-g maneuvers and can reach a maximum speed of over Mach 2.

The Fighting Falcon includes innovations such as a frameless bubble canopy for better visibility, side-mounted control stick to ease control during combat maneuvers, and reclined seat to reduce the effect of g-forces on the pilot. The F-16 has an internal M61 Vulcan cannon in the left wing root and has 11 hardpoints for mounting various missiles, bombs and pods. It was also the first fighter aircraft purpose built to sustain 9-g turns. It has a thrust-to-weight ratio greater than one, providing power to climb and accelerate vertically.

Early models could also be armed with up to six AIM-9 Sidewinder heat-seeking short-range air-to-air missiles (AAM), including a single missile mounted on a dedicated rail launcher on each wingtip. Some variants can also employ the AIM-7 Sparrow medium-range radar-guided AAM, and more recent versions can be equipped with the AIM-120 AMRAAM. It can also carry other AAM; a wide variety of air-to-ground missiles, rockets or bombs; electronic countermeasures (ECM), navigation, targeting or weapons pods; and fuel tanks on eleven hardpoints – six under the wings, two on wingtips and three under the fuselage.

The F-16 design employs a cropped-delta planform incorporating wing-fuselage blending and forebody vortex-control strakes; a fixed-geometry, underslung air intake inlet supplying airflow to the single turbofan jet engine; a conventional tri-plane empennage arrangement with all-moving horizontal "stabilator" tailplanes; a pair of ventral fins beneath the fuselage aft of the wing's trailing edge; a single-piece, bird-proof "bubble" canopy; and a tricycle landing gear configuration with the aft-retracting, steerable nose gear deploying a short distance behind the inlet lip. There is a boom-style aerial refueling receptacle located a short distance behind the rear of the canopy. Split-flap speedbrakes are located at the aft end of the wing-body fairing, and an arrestor hook is mounted underneath the aft fuselage. Another fairing is situated at the base of the vertical tail, beneath the bottom of the rudder, and is used to house various items of equipment such as ECM gear or drag chutes. Several later F-16 models, such as the F-16I variant of the Block 50 aircraft, also have a long dorsal fairing "bulge" that runs along the "spine" of the fuselage from the rear of the cockpit to the tail fairing; these fairings can be used to house additional equipment or fuel.

The air intake was designed to be "far enough forward to allow a gradual bend in the air duct up to the engine face to minimize flow losses and far enough aft so it wouldn't weigh too much or be too draggy or destabilizing."

The F-16 was designed to be relatively inexpensive to build and much simpler to maintain than earlier-generation fighters. The airframe is built with about 80% aviation-grade aluminum alloys, 8% steel, 3% composites, and 1.5% titanium. Control surfaces such as the leading-edge flaps, tailerons, and ventral fins make extensive use of bonded aluminum honeycomb structural elements and graphite epoxy laminate skins. The F-16A had 228 access panels over the entire aircraft, about 80% of which can be reached without work stands. The number of lubrication points, fuel line connections, and replaceable modules was significantly reduced compared to its predecessors.

Although the USAF's LWF program had called for an aircraft structural life of only 4,000 flight hours, and capable of achieving 7.33 g with 80% internal fuel, GD's engineers decided from the start to design the F-16's airframe life to last to 8,000 hours and for 9-g maneuvers on full internal fuel. This proved advantageous when the aircraft's mission was changed from solely air-to-air combat to multi-role operations. Changes over time in actual versus planned operational usage and continued weight growth due to the addition of further systems have required several structural strengthening programs.

Aerodynamic studies in the early 1960s demonstrated that the phenomenon known as "vortex lift" could be beneficially harnessed by the adoption of highly swept wing configurations to reach higher angles of attack through use of the strong leading edge vortex flow off a slender lifting surface. Since the F-16 was being optimized for high agility in air combat, GD's designers chose a slender cropped-delta wing with a leading edge sweep of 40° and a straight trailing edge. To improve its ability to perform in a wide range of maneuvers, a variable-camber wing with a NACA 64A-204 airfoil was selected. The camber is adjusted through the use of leading-edge and trailing edge flaperons linked to a digital flight control system (FCS) that automatically adjusts them throughout the flight envelope. The F-16 has a moderate wing loading, which is lower when fuselage lift is considered.

This vortex lift effect can be increased by the addition of an extension of the leading edge of the wing at its root, the juncture with the fuselage, known as a strake. The strakes act as a sort of additional slender, elongated, short-span, triangular wing running from the actual wing root to a point further forward on the fuselage. Blended fillet-like into the fuselage, including along with the wing root, the strake generates a high-speed vortex that remains attached to the top of the wing as the angle of attack increases, thereby generating additional lift. This allows the aircraft to achieve angles of attack beyond the point at which it would normally stall. The use of strakes also allows a smaller, lower-aspect-ratio wing, which in turn increases roll rates and directional stability, while decreasing aircraft weight. The resulting deeper wingroots also increase structural strength and rigidity, reduce structural weight, and increase internal fuel volume.

The F-16 was the first production fighter aircraft intentionally designed to be slightly aerodynamically unstable. This technique, called "relaxed static stability" (RSS), was incorporated to further enhance the aircraft's maneuver performance. Most aircraft are designed with positive static stability, which induces an aircraft to return to its original attitude following a disturbance. This hampers maneuverability, as the tendency to remain in its current attitude opposes the pilot's effort to maneuver; on the other hand, an aircraft with negative static stability will, in the absence of control input, readily deviate from level and controlled flight. Therefore, an aircraft with negative static stability will be more maneuverable than one that is positively stable. When supersonic, a negatively stable aircraft actually exhibits a more positive-trending (and in the F-16's case, a net positive) static stability due to aerodynamic forces shifting aft between subsonic and supersonic flight. At subsonic speeds the fighter is constantly on the verge of going out of control.

To counter this tendency to depart from controlled flight—and avoid the need for constant minute trimming inputs by the pilot, the F-16 has a quadruplex (four-channel) fly-by-wire (FBW) flight control system (FLCS). The flight control computer (FLCC), which is the key component of the FLCS, accepts the pilot's input from the stick and rudder controls, and manipulates the control surfaces in such a way as to produce the desired result without inducing a loss of control. The FLCC also takes thousands of measurements per second of the aircraft's attitude, and automatically makes corrections to counter deviations from the flight path that were not input by the pilot; coordinated turn is also obtained in such a way that it updates itself by thousands of instructions and produces the required control deflection that comes from dynamics of F-16, thereby allowing for stable flight. This has led to a common aphorism among F-16 pilots: "You don't fly an F-16; it flies you."

The FLCC further incorporates a series of limiters that govern movement in the three main axes based on the jet's current attitude, airspeed and angle of attack, and prevent movement of the control surfaces that would induce an instability such as a slip or skid, or a high angle of attack inducing a stall. The limiters also act to prevent maneuvering that would place more than a 9 g load on the pilot or airframe.

Though the FLCC's limiters work well to limit each axis of movement, it was discovered in early production flight testing that "assaulting" multiple limiters at high angles of attack and low speed can result in angles of attack far exceeding the 25-degree threshold of limiting. This is colloquially referred to as simply "departing". Depending on the attitude of the aircraft, it may settle into a deep stall; a near-freefall at 50° to 60° AOA, either upright or inverted. In this "pocket" of very high AOA, the aircraft's attitude is stable, but being far above stall AOA, the control surfaces do not operate effectively. Further, the pitch limiter of the jet, sensing the high AOA, "freezes" the stabilators in an extreme pitch-up or pitch-down in an attempt to recover. To recover, an override is provided that disables the pitch-limiting, which then allows the pilot to "rock" the aircraft's nose up and down using pitch control, eventually overcoming the 50° threshold and achieving a nose-down attitude which will reduce AOA and allow a return to controlled flight.

Unlike the YF-17 which featured a FBW system with traditional hydromechanical controls serving as a backup, the F-16's designers took the innovative step of eliminating mechanical linkages between the stick and rudder pedals and the aerodynamic control surfaces. The F-16's sole reliance on electronics and wires to relay flight commands, instead of the usual cables and mechanical linkage controls, gained the F-16 the early moniker of "the electric jet". The quadruplex design permits "graceful degradation" in flight control response in that the loss of one channel renders the FLCS a "triplex" system. The FLCC began as an analog system on the A/B variants, but has been supplanted by a digital computer system beginning with the F-16C/D Block 40.

The F-16 program has suffered from controls that were sensitive to static electricity or electrostatic discharge (ESD), including 70–80% of the electronics on the C/D models sensitive to ESD in the early 1980s.

The F-16A/B was originally equipped with the Westinghouse AN/APG-66 fire-control radar. Its slotted planar-array antenna was designed to be sufficiently compact to fit into the F-16's relatively small nose. In uplook mode, the APG-66 uses a low pulse-repetition frequency (PRF) for medium- and high-altitude target detection in a low-clutter environment, and in downlook employs a medium PRF for heavy clutter environments. It has four operating frequencies within the X band, and provides four air-to-air and seven air-to-ground operating modes for combat, even at night or in bad weather. The Block 15's APG-66(V)2 model added a new, more powerful signal processor, higher output power, improved reliability, and increased range in a clutter or jamming environments. The Mid-Life Update (MLU) program further upgrades this to the APG-66(V)2A model, which features higher speed and memory.

The AN/APG-68, an evolution of the APG-66, was introduced with the F-16C/D Block 25. The APG-68 has greater range and resolution, as well as 25 operating modes, including ground-mapping, Doppler beam-sharpening, ground moving target, sea target, and track-while-scan (TWS) for up to 10 targets. The Block 40/42's APG-68(V)1 model added full compatibility with Lockheed Martin Low-Altitude Navigation and Targeting Infra-Red for Night (LANTIRN) pods, and a high-PRF pulse-Doppler track mode to provide continuous-wave (CW) target illumination for semi-active radar-homing (SARH) missiles like the AIM-7 Sparrow. The Block 50/52 F-16s initially received the more reliable APG-68(V)5 which has a programmable signal processor employing Very-High-Speed

Integrated Circuit (VHSIC) technology. The Advanced Block 50/52 (or 50+/52+) are equipped with the APG-68(V)9 radar which has a 30% greater air-to-air detection range, and a synthetic aperture radar (SAR) mode for high-resolution mapping and target detection and recognition. In August 2004, Northrop Grumman received a contract to begin upgrading the APG-68 radars of the Block 40/42/50/52 aircraft to the (V)10 standard, which will provide the F-16 with all-weather autonomous detection and targeting for the use of Global Positioning System (GPS)-aided precision weapons. It also adds SAR mapping and terrain-following (TF) modes, as well as interleaving of all modes.

The F-16E/F is outfitted with Northrop Grumman's AN/APG-80 Active Electronically Scanned Array (AESA) radar, making it only the third fighter to be so equipped. Northrop Grumman is continuing development upon this latest radar, to form the Scalable Agile Beam Radar (SABR). In July 2007, Raytheon announced that it was developing a new Raytheon Next Generation Radar (RANGR) based on its earlier AN/APG-79 AESA radar as an alternative candidate to Northrop Grumman's AN/APG-68 and AN/APG-80 for the F-16.

The powerplant first selected for the single-engined F-16 was the Pratt & Whitney F100-PW-200 afterburning turbofan, a slightly modified version of the F100-PW-100 used by the F-15. Rated at 23,830 lbf (106.0 kN) thrust, it remained the standard F-16 engine through the Block 25, except for new-build Block 15s with the Operational Capability Upgrade (OCU). The OCU introduced the 23,770 lbf (105.7 kN) F100-PW-220, which was also installed on Block 32 and 42 aircraft; the main difference being a Digital Electronic Engine Control (DEEC) unit, which improved engine reliability and reduced the risk of engine stalls. Added to the F-16 production line in 1988, the "-220" also supplanted the F-15's "-100," increasing commonality. Many of the "-220" jet engines on Block 25 and later aircraft were upgraded from mid-1997 to the "-220E" standard, which enhanced reliability and engine maintainability; the changes allowed for a 35% reduction of unscheduled engine removals.

Development of the F100-PW-220/220E was the result of the USAF's Alternate Fighter Engine (AFE) program (colloquially known as "the Great Engine War"), which also saw the entry of General Electric as an F-16 engine provider. Its F110-GE-100 turbofan required modification of the F-16's inlet; the original inlet limited the GE jet's maximum thrust to 25,735 lbf (114.5 kN), while the new Modular Common Inlet Duct allowed the F110 to achieve its maximum thrust of 28,984 lbf (128.9 kN) in afterburner. (To distinguish between aircraft equipped with these two engines and inlets, from the Block 30 series on, blocks ending in "0" (e.g., Block 30) are powered by GE, and blocks ending in "2" (e.g., Block 32) are fitted with Pratt & Whitney engines.)

Further development by these competitors under the Increased Performance Engine (IPE) effort led to the 29,588 lbf (131.6 kN) F110-GE-129 on the Block 50 and 29,160 lbf (129.4 kN) F100-PW-229 on the Block 52. F-16s began flying with these IPE engines on 22 October 1991 and 22 October 1992, respectively. Altogether, of the 1,446 F-16C/Ds ordered by the USAF, 556 were fitted with F100-series engines and 890 with F110s. The United Arab Emirates' Block 60 is powered by the General Electric F110-GE-132 turbofan, which is rated at a maximum thrust of 32,500 lbf (144.6 kN), the highest ever developed for the F-16 aircraft.

Due to their ubiquity, F-16s have participated in numerous conflicts, most of them in the Middle East.

The F-16 is being used by the USAF active, reserve, and Air National Guard units, the USAF aerial demonstration team, the U.S. Air Force Thunderbirds, and as an adversary/aggressor aircraft by the United States Navy.

The U.S. Air Force has flown the F-16 in combat during Operation Desert Storm in 1991, and in the Balkans later in the 1990s. F-16s have patrolled the no fly zones in Iraq during Northern Watch and Southern Watch. They have served during the wars in Afghanistan and Iraq in the 2000s. Most recently, the U.S. has deployed them to enforce the no-fly zone in Libya.

The F-16 is scheduled to remain in service with the U.S. Air Force until 2025. The planned replacement is the Lockheed Martin F-35 Lightning II, which will gradually begin replacing a number of multirole aircraft among the program's member nations.

The F-16's first air-to-air combat success was achieved by the Israeli Air Force (IAF) over the Bekaa Valley on 28 April 1981, against a Syrian Mi-8 helicopter, which was downed with cannon fire. On 7 June 1981, eight Israeli F-16s, escorted by F-15s, executed

Operation Opera, their first employment in a significant air-to-ground operation. This raid severely damaged Osirak, an Iraqi nuclear reactor under construction near Baghdad, to prevent the regime of Saddam Hussein from using the reactor for the creation of nuclear weapons.

The following year, during Operation Peace for Galilee (Lebanon War) Israeli F-16s engaged Syrian aircraft in one of the largest air battles involving jet aircraft, which began on 9 June and continued for two more days. Israeli Air Force F-16s were credited with numerous air-to-air kills during the conflict. F-16s were also used in their ground-attack role for strikes against targets in Lebanon. IAF F-16s participated in the 2006 Lebanon War and during the attacks in the Gaza strip in December 2008.

During the Soviet-Afghan war, between May 1986 and January 1989, Pakistan Air Force (PAF) F-16s shot down at least 10 intruders from Afghanistan.

The Pakistan Air Force has used its F-16s in various foreign and internal military exercises, such as the "Indus Vipers" exercise in 2008 conducted jointly with Turkey. Since May 2009, the PAF has also been using their F-16 fleet to attack militant positions and support the Pakistan Army's operations in North-West Pakistan against the Taliban insurgency.

The Royal Netherlands Air Force, Belgian Air Force, Turkish Air Force, Royal Danish Air Force, and Royal Norwegian Air Force, and Venezuela have flown the F-16 on combat missions. A Dutch F-16AM shot down a Serbian MiG-29 during the Kosovo War in 1999. Belgian and Danish F-16s also supported operations in Kosovo.

ION THRUSTER

An ion thruster (fig. 89) is a form of electric propulsion used for spacecraft propulsion that creates thrust by accelerating ions. Ion thrusters are categorized by how they accelerate the ions, using either electrostatic or electromagnetic force. Electrostatic ion thrusters use the Coulomb force and accelerate the ions in the direction of the electric field. Electromagnetic ion thrusters use the Lorentz force to accelerate the ions. The term "ion thruster" by itself usually denotes the electrostatic or gridded ion thrusters.

Fig. 89. NASA's 2.3 kW NSTAR ion thruster during a hot fire test at the Jet Propulsion Laboratory on the Deep Space 1 spacecraft.

The thrust created in ion thrusters is very small compared to conventional chemical rockets, but a very high specific impulse, or propellant efficiency, is obtained. This high propellant efficiency is achieved through the very frugal propellant consumption of the ion thruster propulsion system. They do, however, use a large amount of power, and in use their performance is power limited (whereas normal chemical thrusters are energy limited). Given the practical weight of suitable power sources, the accelerations given by these types of thrusters is of the order of one thousandth of standard gravity.

Due to their relatively high power needs, given the specific power of power supplies, and the requirement of an environment void of other ionized particles, ion thrust propulsion is currently only practical beyond planetary atmosphere (in space).

The official father of the concept of electric propulsion is Konstantin Tsiolkovsky as he is the first to publish mention of the idea in 1911. However, the first documented instance where the possibility of electric propulsion is considered is found in Robert H.

Goddard's handwritten notebook in an entry dated 6 September 1906. The first experiments with ion thrusters were carried out by Goddard at Clark University from 1916–1917. The technique was recommended for near-vacuum conditions at high altitude, but thrust was demonstrated with ionized air streams at atmospheric pressure. The idea appeared again in Hermann Oberth's "Wege zur Raumschiffahrt" (Ways to Spaceflight), published in 1923, where he explained his thoughts on the mass savings of electric propulsion, predicted its use in spacecraft propulsion and attitude control, and advocated electrostatic acceleration of charged gases.

A working ion thruster was built by Harold R. Kaufman in 1959 at the NASA Glenn Research Center facilities. It was similar to the general design of a gridded electrostatic ion thruster with mercury as its fuel. Suborbital tests of the engine followed during the 1960s and in 1964 the engine was sent into a suborbital flight aboard the Space Electric Rocket Test 1 (SERT 1). It successfully operated for the planned 31 minutes before falling back to Earth.

The Hall effect thruster was studied independently in the U.S. and the USSR in the 1950s and 60s (fig. 90). However, the concept of a Hall thruster was only developed into an efficient propulsion device in the former Soviet Union, whereas in the U.S., scientists focused instead on developing gridded ion thrusters. Hall effect thrusters were operated on Soviet satellites since 1972. Until the 1990s they were mainly used for satellite stabilization in North-South and in East-West directions. Some 100-200 engines completed their mission on Soviet and Russian satellites until the late 1990s. Soviet thruster design was introduced to the West in 1992 after a team of electric propulsion specialists, under the support of the Ballistic Missile Defense Organization, visited Soviet laboratories.

Fig. 90. Soviet and Russian Hall effect thrusters.

Ion thrusters use beams of ions (electrically charged atoms or molecules) to create thrust in accordance with momentum conservation.

The method of accelerating the ions varies, but all designs take advantage of the charge/mass ratio of the ions.

This ratio means that relatively small potential differences can create very high exhaust velocities.

This reduces the amount of reaction mass or fuel required, but increases the amount of specific power required compared to chemical rockets.

Ion thrusters are therefore able to achieve extremely high specific impulses.

The drawback of the low thrust is low spacecraft acceleration because the mass of current electric power units is directly correlated with the amount of power given.

This low thrust makes ion thrusters unsuited for launching spacecraft into orbit, but they are ideal for in-space propulsion applications.

Various ion thrusters have been designed and they all generally fit under two categories.

The thrusters are categorized as either electrostatic or electromagnetic.

The main difference is how the ions are accelerated.

Electrostatic ion thrusters use the Coulomb force and are categorized as accelerating the ions in the direction of the electric field.

Electromagnetic ion thrusters use the Lorentz force to accelerate the ions.

Power supplies for ion thrusters are usually solar panels, but at sufficiently large distances from the Sun, nuclear power is used.

In each case the power supply mass is essentially proportional to the peak power that can be supplied, and they both essentially give, for this application, no limit to the energy.

Electrostatic ion thrusters

Gridded electrostatic ion thrusters

Gridded electrostatic ion thrusters commonly utilize xenon gas (fig. 91).

This gas has no charge and is ionized by bombarding it with energetic electrons.

These electrons can be provided from a hot cathode filament and accelerated in the electrical field of the cathode fall to the anode (Kaufman type ion thruster).

Alternatively, the electrons can be accelerated by the oscillating electric field induced by an alternating magnetic field of a coil, which results in a self-sustaining discharge and omits any cathode (radiofrequency ion thruster).

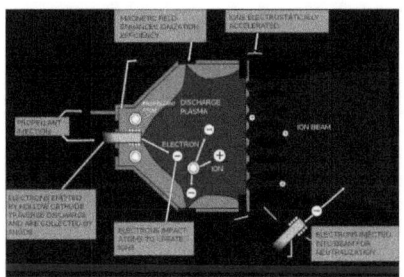

Fig. 91. *A diagram of how a gridded electrostatic ion engine (Kaufman type) works.*

The positively charged ions are extracted by an extraction system consisting of 2 or 3 multi-aperture grids.

After entering the grid system via the plasma sheath the ions are accelerated due to the potential difference between the first and second grid (named screen and accelerator grid) to the final ion energy of typically 1-2 keV, thereby generating the thrust.

Ion thrusters emit a beam of positive charged xenon ions only.

To avoid charging-up the spacecraft, another cathode is placed near the engine, which emits electrons (basically the electron current is the same as the ion current) into the ion beam.

This also prevents the beam of ions from returning to the spacecraft and thereby cancelling the thrust.

Gridded electrostatic ion thruster research (past/present):

NASA Solar electric propulsion Technology Application Readiness (NSTAR);

NASA's Evolutionary Xenon Thruster (NEXT);

Nuclear Electric Xenon Ion System (NEXIS);

High Power Electric Propulsion (HiPEP);

EADS Radio-Frequency Ion Thruster (RIT);

Dual-Stage 4-Grid (DS4G).

Hall effect thrusters

Hall effect thrusters (fig. 92) accelerate ions with the use of an electric potential maintained between a cylindrical anode and a negatively charged plasma that forms the cathode.

The bulk of the propellant (typically xenon gas) is introduced near the anode, where it becomes ionized, and the ions are attracted towards the cathode, they accelerate towards and through it, picking up electrons as they leave to neutralize the beam and leave the thruster at high velocity.

The anode is at one end of a cylindrical tube, and in the center is a spike that is wound to produce a radial magnetic field between it and the surrounding tube.

The ions are largely unaffected by the magnetic field, since they are too massive.

However, the electrons produced near the end of the spike to create the cathode are far more affected and are trapped by the magnetic field, and held in place by their attraction to the anode. Some of the electrons spiral down towards the anode, circulating around the spike in a Hall current.

When they reach the anode they impact the uncharged propellant and cause it to be ionized, before finally reaching the anode and closing the circuit.

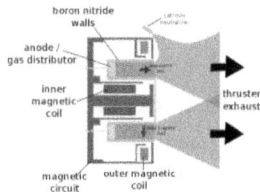

Fig. 92. *Schematic of a Hall Thruster.*

Field emission electric propulsion

Field emission electric propulsion (FEEP) thrusters use a very simple system of accelerating liquid metal ions to create thrust. Most designs use either caesium or indium as the propellant.

The design consists of a small propellant reservoir that stores the liquid metal, a very small slit that the liquid flows through, and then the accelerator ring. Caesium and indium are used due to their high atomic weights, low ionization potentials, and low melting points. Once the liquid metal reaches the inside of the slit in the emitter, an electric field applied between the emitter and the accelerator ring causes the liquid metal to become unstable and ionize. This creates a positive ion, which can then be accelerated in the electric field created by the emitter and the accelerator ring. These positively charged ions are then neutralized by an external source of electrons to prevent charging of the spacecraft hull.

Electromagnetic thrusters

Pulsed inductive thrusters (PIT)

Pulsed inductive thrusters (PIT) use pulses of thrust instead of one continuous thrust, and have the ability to run on power levels in the order of Megawatts (MW).

PITs consist of a large coil encircling a cone shaped tube that emits the propellant gas as shown in the diagram. Ammonia is the gas commonly used in PIT engines. For each pulse of thrust the PIT gives, a large charge first builds up in a group of capacitors behind the coil and is then released. This creates a current that moves circularly in the direction of $j\theta$.

The current then creates a magnetic field in the outward radial direction (Br), which then creates a current in the ammonia gas that has just been released in the opposite direction of the original current. This opposite current ionizes the ammonia and these positively charged ions are accelerated away from the PIT engine due to the electric field $j\theta$ crossing with the magnetic field Br, which is due to the Lorentz Force.

Magnetoplasmadynamic(MPD)/lithium Lorentz force accelerator (LiLFA)

Magnetoplasmadynamic (MPD) thrusters and lithium Lorentz force accelerator (LiLFA) thrusters use roughly the same idea with the LiLFA thruster building off of the MPD thruster.

Hydrogen, argon, ammonia, and nitrogen gas can be used as propellant. In a certain configuration, the ambient gas in Low Earth Orbit (LEO) can be used as a propellant.

The gas first enters the main chamber where it is ionized into plasma by the electric field between the anode and the cathode. This plasma then conducts electricity between the anode and the cathode. This new current creates a magnetic field around the cathode, which crosses with the electric field, thereby accelerating the plasma due to the Lorentz Force.

The LiLFA thruster uses the same general idea as the MPD thruster, except for two main differences. The first difference is that the LiLFA uses lithium vapor, which has the advantage of being able to be stored as a solid.

The other difference is that the cathode is replaced by multiple smaller cathode rods packed into a hollow cathode tube. The cathode in the MPD thruster is easily corroded due to constant contact with the plasma. In the LiLFA thruster the lithium vapor is injected into the hollow cathode and is not ionized to its plasma form/corrode the cathode rods until it exits the tube. The plasma is then accelerated using the same Lorentz Force.

The Magnetoplasmadynamic (MPD, fig. 93) thruster (MPDT) is a form of electrically powered spacecraft propulsion which uses the Lorentz force (a force resulting from the interaction between a magnetic field and an electric current) to generate thrust. It is sometimes referred to as Lorentz Force Accelerator (LFA) or (mostly in Japan) MPD arcjet.

Fig. 93. An MPD thruster during test firing.

Fig. 94. CGI rendering of Princeton self-field MPD thruster (from Popular Mechanics magazine)

Generally, a gaseous fuel is ionized and fed into an acceleration chamber, where the magnetic and electrical fields are created using a power source.

The particles are then propelled by the Lorentz force resulting from the interaction between the current flowing through the plasma and the magnetic field (which is either externally applied, or induced by the current) out through the exhaust chamber. Unlike chemical propulsion, there is no combustion of fuel. As with other electric propulsion variations, both specific impulse and thrust increase with power input, while thrust per watt drops.

There are two main types of MPD thrusters, applied-field and self-field (fig. 94). Applied-field thrusters have magnetic rings surrounding the exhaust chamber to produce the magnetic field, while self-field thrusters have a cathode extending through the middle of the chamber. Applied fields are necessary at lower power levels, where self-field configurations are too weak. Various propellants such as xenon, neon, argon, hydrazine, and lithium have been used, with lithium generally being the best performer.

The VASIMR is a totally different type of engine that attempts to provide the same level of performance as MPD but operates on a totally different principles : it is an electrothermal device, where the energy is first applied to the propellant in order to increase its random kinetic energy (temperature), in case of VASIMR the propellant is heated using RF and then a part of the thermal energy content of the propellant is converted into directed kinetic energy by using an appropriate nozzle, in this case a magnetic nozzle. Details on this engine can be found in the main Variable Specific Impulse Magnetoplasma Rocket article.

In theory, MPD thrusters could produce extremely high specific impulses (Isp) with an exhaust velocity of up to and beyond 110,000 m/s, triple the value of current xenon-based ion thrusters, and about 20 times better than liquid rockets.

MPD technology also has the potential for thrust levels of up to 200 newtons (N) (45 lbf), by far the highest for any form of electric propulsion, and nearly as high as many interplanetary chemical rockets.

This would allow use of electric propulsion on missions which require quick delta-v maneuvers (such as capturing into orbit around another planet), but with many times greater fuel efficiency.

MPD thruster technology has been explored academically, but commercial interest has been low due to several remaining problems. One big problem is that power requirements on the order of hundreds of kilowatts are required for optimum performance. Current interplanetary spacecraft power systems (such as radioisotope thermoelectric generators (RTGs)) and solar arrays are incapable of producing that much power.

NASA's Project Prometheus reactor was expected to generate power in the hundreds of kilowatts range but was discontinued in 2005.

A project to produce a space-going nuclear reactor designed to generate 600 kilowatts of electrical power began in 1963 and ran for most of the 1960s in the USSR.

It was to power a communication satellite which was in the end not approved.

Nuclear reactors supplying kilowatts of electrical power (of the order of ten times more than current RTG power supplies) have been orbited by the USSR: RORSAT and TOPAZ.

Plans to develop a megawatt-scale nuclear reactor for the use aboard a manned spaceship were announced in 2009 by Russian nuclear Kurchatov Institute, national space agency Roskosmos, and confirmed by the President of Russia in November 2009 address.

Another plan, proposed by Bradley C. Edwards, is to beam power from the ground.

This plan utilizes 5 200 kW free electron lasers at 0.84 micrometres with adaptive optics on the ground to beam power to the MPD-powered spacecraft, where it is converted to electricity by GaAs photovoltaic panels.

The tuning of the laser wavelength of 0.840 micrometres (1.48 eV per photon) and the PV panel bandgap of 1.43 eV to each other produces an estimated conversion efficiency of 59% and a predicted power density of up to 540 kW/m2.

This would be sufficient to power a MPD upper stage, perhaps to lift satellites from LEO to GEO.

Other problem with MPD technology has been the degradation of cathodes due to evaporation driven by high current densities (in excess of 100 amps/cm^2).

The use of lithium and barium propellant mixtures and multi-channel hollow cathodes has been shown in the laboratory to be a promising solution for the cathode erosion problem.

The Variable Specific Impulse Magnetoplasma Rocket (VASIMR) is an electro-magnetic thruster for spacecraft propulsion. It uses radio waves to ionize and heat a propellant and magnetic fields to accelerate the resulting plasma to generate thrust. It is one of several types of spacecraft electric propulsion systems.

The method of heating plasma used in VASIMR was originally developed as a result of research into nuclear fusion. VASIMR is intended to bridge the gap between high-thrust, low-specific impulse propulsion systems and low-thrust, high-specific impulse systems.

VASIMR is capable of functioning in either mode. Costa Rican scientist and former astronaut Franklin Chang-Diaz created the

VASIMR concept and has been working on its development since 1977 (fig. 95).

Fig. 95. *Several VASIMR engines propelling a craft through space.*

The Variable Specific Impulse Magnetoplasma Rocket, sometimes referred to as the Electro-thermal Plasma Thruster or Electro-thermal Magnetoplasma Rocket, uses radio waves to ionize and heat propellant and magnetic fields, accelerating the resulting plasma which generates thrust.

This type of engine is electrodeless and as such belongs to the same electric propulsion family (while differing in the method of plasma acceleration) as the electrodeless plasma thruster, the microwave arcjet, or the pulsed inductive thruster class.

It can also be seen as an electrodeless version of an arcjet, able to reach higher propellant temperature by limiting the heat flux from the plasma to the structure. Neither type of engine has any electrodes.

The main advantage of such designs is elimination of problems with electrode erosion that cause rival designs of ion thrusters which use electrodes to have a short life expectancy. Furthermore, since every part of a VASIMR engine is magnetically shielded and does not come into direct contact with plasma, the potential durability of this engine design is greater than other ion/plasma engine designs.

The engine design encompasses three parts: turning gas into plasma via helicon RF antennas; energizing plasma via further RF heating in an ion cyclotron resonance frequency (ICRF) booster; and using electromagnets to create a magnetic nozzle to convert the plasma's built-up thermal energy into kinetic force (fig. 96).

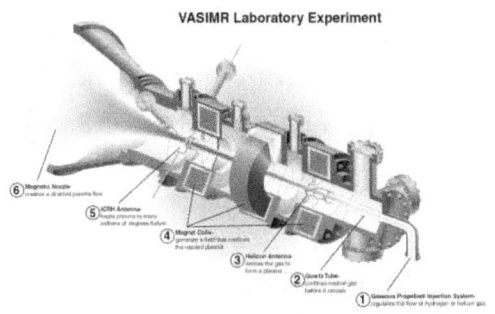

Fig. 96. *Several VASIMR engines propelling a craft through space.*

By varying the amount of energy dedicated to RF heating and the amount of propellant delivered for plasma generation VASIMR is capable of either generating low-thrust, high-specific impulse exhaust or relatively high-thrust, low-specific impulse exhaust.

In contrast with usual cyclotron resonance heating processes, in VASIMR, ions are immediately ejected through the magnetic nozzle before they have time to achieve thermalized distribution.

Based on novel theoretical work in 2004 by Arefiev and Breizman of UT-Austin, virtually all of the energy in the ion cyclotron wave is uniformly transferred to ionized plasma in a single-pass cyclotron absorption process.

This allows for ions to leave the magnetic nozzle with a very narrow energy distribution and for significantly simplified and compact magnet arrangement in the engine.

VASIMR does not use electrodes and magnetically shields plasma from most of the hardware parts, thus eliminating electrode erosion, a major source of wear and tear in ion engines.

Compared to traditional rocket engines with very complex plumbing, high performance valves, actuators and turbopumps, VASIMR eliminates practically all moving parts from its design (apart from minor ones like gas valves), maximizing its long term durability.

However, some new problems emerge like interaction with strong magnetic fields and thermal management.

The relatively large power at which VASIMR operates generates a lot of waste heat which needs to be channeled away without creating thermal overload and undue thermal stress on materials used.

Powerful superconducting electromagnets, employed to contain hot plasma, generate tesla-range magnetic fields.

They can present problems with other on board devices and also can adversely interact with Earth magnetosphere.

To counter this latter effect the VF-200 will consist of two 100 kW thruster units packaged together with the magnetic field of each thruster oriented in opposite directions in order to make a zero-torque magnetic quadrapole.

After many years researching the concept with NASA, Franklin Chang-Diaz set up the Ad Astra Rocket Company in January 2005 to begin development of the VASIMR engine.

Later that year, the company signed a Space Act Agreement with NASA, and were granted control of the Advanced Space Propulsion Laboratory.

In this lab, a 50 kW prototype was constructed, and underwent testing in a vacuum chamber. Later, a 100 kW version was developed, and this was followed by a 200 kW prototype. After a long period of rigorous testing in a 150 m3 vacuum chamber, the latest configuration was deemed space-worthy, and it was announced that the company had entered into an agreement to test the engine on the International Space Station, in or before 2013.

The first VASIMR engine model VX-50 proved to be capable of 0.5 newtons (0.1 lbf) thrust.

Published data on the VX-50 engine, capable of processing 50 kW of total radio frequency power, showed thruster efficiency to be 59% calculated as: 90% NA coupling efficiency × 65% NB ion speed boosting efficiency.

It was hoped that the overall efficiency of the engine could be increased by scaling up power levels.

Model VX-100 was expected to have a thruster efficiency of 72% by improving the NB ion speed boosting efficiency to 80%.

There were, however, additional (smaller) efficiency losses related to the conversion of DC electric current to radio frequency power and also to the superconducting magnets' auxiliary equipment energy consumption.

By comparison, 2009 state-of-the-art, proven ion engine designs such as NASA's HiPEP operated at 80% total thruster/PPU energy efficiency.

On October 24, 2008 the company announced that the plasma generation aspect of the VX-200 engine - helicon first stage or solid-state high frequency power transmitter - had reached operational status.

The key enabling technology, solid-state DC-RF power-processing, has become very efficient reaching up to 98% efficiency.

The helicon discharge uses 30 kWe of radio waves to turn argon gas into plasma.

The remaining 170 kWe of power is allocated for passing energy to, and acceleration of, plasma in the second part of the engine via ion cyclotron resonance heating.

Based on data released from previous VX-100 testing, it was expected that the VX-200 engine would have a system efficiency of 60-65% and thrust level of 5N.

Optimal specific impulse appeared to be around 5000s using low cost argon propellant.

The specific power estimated at 1.5 kg/kW meant that this version of the VASIMR engine would weigh only about 300 kg.

One of the remaining untested issues was potential vs actual thrust; that is, whether the hot plasma actually detached from the rocket. Another issue was waste heat management (60% efficiency means about (100%-60%)/100%*200 kW = 80 kW of unnecessary heat) critical to allowing for continuous operation of VASIMR engine.

Between April and September 2009, tests were performed on the VX-200 prototype with fully integrated 2 Tesla superconducting magnets.

They successfully expanded the power range of the VASIMR up to its full operational capability of 200 kW.

During November 2010, long duration, full power firing tests were performed with the VX-200 engine reaching the steady state operation for 25 seconds thus validating basic design characteristics.

Results presented to NASA and academia in January 2011 have confirmed that the design point for optimal efficiency on the VX-200 is 50 km/s exhaust velocity, or an Isp of 5000 s.

Based on these data, thruster efficiency of 70% has been deemed by Ad Astra to be achievable, yielding an overall system efficiency (DC electricity to thruster power) of 60% (since the DC to RF power conversion efficiency exceeds 95%).

On December 8, 2008, Ad Astra Company signed an agreement with NASA to arrange the placement and testing of a flight version of the VASIMR, the VF-200, on the International Space Station (ISS).

As of February 2011, its launch is anticipated to be in 2014, though it may be later.

The Taurus II has been reported as the "top contender" for the launch vehicle.

Since the available power from the ISS is less than 200 kW, the ISS VASIMR will include a trickle-charged battery system allowing for 15 min pulses of thrust.

Testing of the engine on ISS is valuable because it orbits at a relatively low altitude and experiences fairly high levels of atmospheric drag, making periodic boosts of altitude necessary. Currently, altitude reboosting by chemical rockets fulfills this requirement.

If the tests of VASIMR reboosting of the ISS goes according to plan, the increase in specific impulse could mean that the cost of fuel for altitude reboosting will be one-twentieth of the current $210 million annual cost.

Hydrogen is generated by the ISS as a by-product, which is currently vented into space, it will be redirected to the VASMIR to act as the fuel in place of the current Argon.

The VF-200 flight-rated thruster consists of two 100 kW VASIMR units with opposite magnetic dipoles so that no net rotational torque is applied to the space station when the thrusters are firing. The VF-200-1 is the first flight unit and will be tested in space attached to the ISS (fig. 97).

As of February 2011, NASA has 100 people assigned to the project to work with Ad Astra to integrate the VF-200 onto the space station.

VASIMR is not suitable to launch payloads from the surface of the Earth due to its low thrust-to-weight ratio and its need of a vacuum to operate.

Instead, it would function as an upper stage for cargo, reducing the fuel requirements for in-space transportation. The engine is expected to perform the following functions at a fraction of the cost of chemical technologies:

> drag compensation for space stations
>
> lunar cargo delivery
>
> satellite repositioning
>
> satellite refueling, maintenance and repair
>
> in space resource recovery
>
> ultra fast deep space robotic missions

Other applications for VASIMR such as the rapid transportation of people to Mars would require a very high power, low mass energy source, such as a nuclear reactor (see nuclear electric rocket).

NASA Administrator Charles Bolden said that VASIMR technology could be the breakthrough technology that would reduce the travel time on a Mars mission from months to days.

In August 2008, Tim Glover, Ad Astra director of development, publicly stated that the first expected application of VASIMR engine is "hauling things [non-human cargo] from low-Earth orbit to low-lunar orbit" supporting NASA's return to Moon efforts.

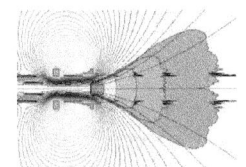

Fig. 97. *VASIMR magnetic field.*

The most important near-future application of VASIMR-powered spacecraft is transportation of cargo.

Numerous studies have shown that, despite longer transit times, VASIMR-powered spacecraft will be much more efficient than traditional integrated chemical rockets at moving goods through space.

An orbital transfer vehicle (OTV) — essentially a "space tug" — powered by a single VF-200 engine would be capable of transporting about 7 metric tons of cargo from low Earth orbit (LEO) to low Lunar orbit (LLO) with about a six month transit time.

NASA envisages delivering about 34 metric tons of useful cargo to LLO in a single flight with a chemically propelled vehicle.

To make that trip, about 60 metric tons of LOX-LH2 propellant would be burned.

A comparable OTV would need to employ 5 VF-200 engines powered by a 1 MW solar array.

To do the same job, such OTV would need to expend only about 8 metric tons of argon propellant.

Total mass of such electric OTV would be in the range of 49 t (outbound & return fuel: 9 t, hardware: 6 t, cargo 34 t). The OTV transit times can be reduced by carrying lighter loads and/or expending more argon propellant with VASIMR throttled down to lower Isp.

For instance, an empty OTV on the return trip to Earth covers the distance in about 23 days at optimal specific impulse of 5,000 s (50 kN·s/kg) or in about 14 days at Isp of 3,000 s (30 kN·s/kg).

The total mass of the NASA specs' OTV (including structure, solar array, fuel tank, avionics, propellant and cargo) was assumed to be 100 metric tons (98.4 long tons; 110 short tons) allowing almost double the cargo capacity compared to chemically propelled vehicle but requiring even bigger solar arrays (or other source of power) capable of providing 2 MW.

As of October 2010, Ad Astra Rocket Company is working toward utilizing VASIMR technology for space tug missions to help "clean up the ever-growing problem of space trash." They hope to have a first-generation commercial offering by 2014.

Electrodeless plasma thrusters

Electrodeless plasma thrusters have two unique features: the removal of the anode and cathode electrodes and the ability to throttle the engine. The removal of the electrodes takes away the factor of erosion, which limits lifetime on other ion engines. Neutral gas is first ionized by electromagnetic waves and then transferred to another chamber where it is accelerated by an oscillating electric and magnetic field, also known as the ponderomotive force. This separation of the ionization and acceleration stage give the engine the ability to throttle the speed of propellant flow, which then changes the thrust magnitude and specific impulse values.

Electrothermal thrusters

Electrothermal thrusters use electric power to accelerate propellant. There are several types:

1. Resistojet

2. Arcjet

3. Microwave electrothermal thrusters

4. Ion Cyclotron Heating thrusters (VASIMR)

Helicon double layer thruster

A helicon double layer thruster is a type of plasma thruster, which ejects high velocity ionized gas to provide thrust to a spacecraft.

In this thruster design, gas is injected into a tubular chamber (the source tube) with one open end. Radio frequency AC power (at 13.56 MHz in the prototype design) is coupled into a specially shaped antenna wrapped around the chamber.

The electromagnetic wave emitted by the antenna causes the gas to break down and form a plasma. The antenna then excites a Helicon wave in the plasma, which further heats the plasma. The device has a roughly constant magnetic field in the source tube (supplied by Solenoids in the prototype), but the magnetic field diverges and rapidly decreases in magnitude away from the source region, and might be thought of as a kind of magnetic nozzle.

In operation, there is a sharp boundary between the high density plasma inside the source region, and the low density plasma in the exhaust, which is associated with a sharp change in electrical potential. The plasma properties change rapidly across this boundary, which is known as a current free electric double layer.

The electrical potential is much higher inside the source region than in the exhaust, and this serves both to confine most of the electrons, and to accelerate the ions away from the source region. Enough electrons escape the source region to ensure that the plasma in the exhaust is neutral overall.

A major limiting factor of ion thrusters is their small thrust; however, it is generated at a high propellant efficiency (mass utilisation, specific impulse). The efficiency comes from the high exhaust velocity, which in turn demands high energy, and the performance is ultimately limited by the available spacecraft power.

The low thrust requires ion thrusters to provide continuous thrust for a long time to achieve the needed change in velocity (delta-v) for a particular mission. To cause enough change in momentum, ion thrusters are designed to last for periods of weeks to years.

In practice the lifetime of electrostatic ion thrusters is limited by several processes:

In electrostatic gridded ion thruster design, charge-exchange ions produced by the beam ions with the neutral gas flow can be accelerated towards the negatively biased accelerator grid and cause grid erosion.

End-of-life is reached when either a structural failure of the grid occurs or the holes in the accelerator grid become so large that the ion extraction is largely affected (e.g. by the occurrence of electron backstreaming).

Grid erosion cannot be avoided and is the major lifetime-limiting factor. By a thorough grid design and material selection, lifetimes of 20,000 hours and far beyond are reached, which is sufficient to fulfill current space missions.

A test of the NASA Solar electric propulsion Technology Application Readiness (NSTAR) electrostatic ion thruster resulted in 30,472 hours (roughly 3.5 years) of continuous thrust at maximum power.

The test was concluded prior to any failure and test results showed the engine was not approaching failure either.

Hall thrusters suffer from very strong erosion of the ceramic discharge chamber.

Due to the rather high discharge voltages of up to 1000V energetic ions can impinge to the chamber walls and erode material. Lifetimes of a few thousand hours are reached.

Ionization energy represents a very large percentage of the energy needed to run ion drives.

The ideal propellant for ion drives is thus a propellant molecule or atom that is easy to ionise, that has a high mass/ionisation energy ratio.

In addition, the propellant should not cause erosion of the thruster to any great degree to permit long life; and should not contaminate the vehicle.

Many current designs use xenon gas as it is easy to ionize, has a reasonably high atomic number, its inert nature, and low erosion. However, xenon is globally in short supply and very expensive.

Older designs used mercury, but this is toxic and expensive, tended to contaminate the vehicle with the metal and was difficult to feed accurately.

Other propellants such as bismuth show promise and are areas of research, particularly for gridless designs such as Hall effect thrusters.

VASIMR design (and other plasma based engines) are theoretically able to use practically any material for propellant. However, in current tests the most practical propellant is argon, which is a relatively abundant and cheap gas.

Ion thrusters are frequently quoted with an efficiency metric. This efficiency is the kinetic energy of the exhaust jet emitted per second divided by the electrical power into the device.

The actual overall system energy efficiency in use is determined by the propulsive efficiency, which depends on vehicle speed and exhaust speed.

Some thrusters can vary exhaust speed in operation, but all can be designed with different exhaust speeds.

At the lower end of Isps the overall efficiency drops because the ionisation takes up a larger percentage energy, and at the high end propulsive efficiency is reduced.

Optimal efficiencies and exhaust velocities can thus be calculated for any given mission to give minimum overall cost.

Ion thrusters have many applications for in-space propulsion. The best applications of the thrusters make use of the long lifetime when significant thrust is not needed.

Examples of this include orbit transfers, attitude adjustments, drag compensation for low earth orbits, transporting cargo such as chemical fuels between propellant depots and ultra fine adjustments for more scientific missions.

Ion thrusters can also be used for interplanetary and deep space missions where time is not crucial. Continuous thrust over a very long time can build up a larger velocity than traditional chemical rockets.

Of all the electric thrusters, ion thrusters have been the most seriously considered commercially and academically in the quest for interplanetary missions and orbit raising maneuvers. Ion thrusters are seen as the best solution for these missions as they require very high change in velocity overall that can be built up over long periods of time.

Several spacecraft have operated with this technology.

SERT

Ion propulsion systems were first demonstrated in space by the NASA Lewis (now Glenn Research Center) missions "Space Electric Rocket Test" (SERT) I and II. The first was SERT-1, launched July 20, 1964, successfully proved that the technology operated as predicted in space. These were electrostatic ion thrusters using mercury and cesium as the reaction mass The second test, SERT-II, launched on February 3, 1970, verified the operation of two mercury ion engines for thousands of running hours.

Deep Space 1

NASA has developed an ion thruster called NSTAR for use in their interplanetary science missions beginning in the late-1990s. This xenon-propelled ion thruster was first space-tested in the highly successful space probe Deep Space 1, launched in 1998. This was the first use of electric propulsion as the interplanetary propulsion system on a science mission. Based on this NASA-developed technology, the contractor, Hughes, developed the XIPS (Xenon Ion Propulsion System) for performing station keeping on geosynchronous satellites.

Artemis

On 12 July 2001, the European Space Agency failed to launch their Artemis telecommunication satellite to desired altitude, and left it in a decaying orbit.

The satellite's chemical propellant supply was sufficient to transfer it to a semi-stable orbit, and over the next 18 months the experimental onboard ion propulsion system RIT-10 (intended only for secondary stationkeeping and maneuvering) was utilized to transfer it to a geostationary orbit.

Hayabusa

The Japanese space agency's Hayabusa, which was launched in 2003 and successfully rendezvoused with the asteroid 25143 Itokawa and remained in close proximity for many months to collect samples and information, was powered by four xenon ion engines. It used xenon ions generated by microwave electron cyclotron resonance, and a carbon / carbon-composite material (which is resistant to erosion) for its acceleration grid.

Although the ion engines on Hayabusa had some technical difficulties, in-flight reconfiguration allowed one of the four engines to be repaired, and allowed the mission to successfully return to Earth.

Smart 1

The Hall effect thruster is a type of ion thruster that has been used for decades for station keeping by the Soviet Union and is now also applied in the West: the European Space Agency's satellite Smart 1, launched in 2003, used it (Snecma PPS-1350-G). This satellite completed its mission on 3 September 2006, in a controlled collision on the Moon's surface, after a trajectory deviation so scientist could see the 3 meter crater the impact created on the visible side of the moon.

GOCE

ESA's Gravity Field and Steady-State Ocean Circulation Explorer was launched on March 16, 2009. It will continue to use ion propulsion throughout its twenty month mission to combat the air-drag it experiences in its low orbit.

BepiColombo

ESA will launch the BepiColombo mission to Mercury in 2014. It uses ion thrusters in combination with swing-bys to get to Mercury, where a chemical rocket will be fired for orbit insertion.

LISA Pathfinder

LISA Pathfinder is an ESA spacecraft to be launched in 2013. It will not use ion thrusters as its primary propulsion system, but will use both colloid thrusters and FEEP for very precise altitude control—the low thrusts of these propulsion devices make it possible to move the spacecraft incremental distances very accurately. It is a test for the possible LISA mission.

International Space Station

As of February 2010, a 2011/2012 launch of an Ad Astra VF-200 200 kW VASIMR electromagnetic thruster is planned for placement and testing on the International Space Station. The VF-200 is a flight version of the VX-200 (though it may be later). Since the available power from the ISS is less than 200 kW, the ISS VASIMR will include a trickle-charged battery system allowing for 15 min pulses of thrust. Testing of the engine on ISS is valuable because ISS orbits at a relatively low altitude and experiences fairly high levels of atmospheric drag, making periodic boosts of altitude necessary. Currently, altitude reboosting by chemical rockets fulfills this requirement. If the tests of VASIMR reboosting of the ISS goes according to plan, the increase in specific impulse could mean that the cost of fuel for altitude reboosting will be one-twentieth of the current $210 million annual cost. Hydrogen is generated by the ISS as a by-product, which is currently vented into space.

BEAM OR IONIC ENGINES

By this chapter the authors propose a new pulse engine which works with beam or ionic (ionic beam) pulses [36].

With a new ionic engine one builds a new aircraft (a new ship).

The principal characteristic of this kind of engine is the high power (energy) which accelerates the beam at very high energy, in circular accelerators, in modern linear accelerators (LINAC), or in both.

One can use accelerators similar with the static physics accelerators (synchrotron, synchrocyclotron or isochronous cyclotron).

Ionic engine (ion thruster, which accelerates the positive ions through a potential difference) is about 10 times more effective than classic system based on combustion.

We can still improve the efficiency of 10-50 times if one uses positive ions accelerated in a cyclotron mounted on the ship; the efficiency can easily grow for 1000 times if the positive ions will be accelerated in a high energy synchrotron, synchrocyclotron or isochronous cyclotron (1-100 GeV) [36].

Future (ionic) engine will have mandatory a circular particle accelerator (high or very high energy; see the figure 99). Sure that the difficulties will arise from design, but they need to be resolved step by step.

We can thus increase the speed and autonomy of the ship using a less quantity of fuel.

It can use synchrotron radiation (synchrotron light, high intensity beams), like high intensity (X-ray or Gamma ray) radiation, as well. In this case will be a beam engine (not an ionic engine).

A linear particle accelerator (also called a LINAC) is an electrical device for the acceleration of subatomic particles. This sort of particle accelerator has many applications.

It used recently as to an injector into a higher energy synchrotron at a dedicated experimental particle physics laboratory. In this, the big classic synchrotron is reduced to a ring surface (magnetic core).

The design of a LINAC depends on the type of particle that is being accelerated: electron, proton or ion.

It proposes using a powerful LINAC at the exit of synchrotron (especially when one accelerates electrons) to not lose energy by photons premature emission (figure 99) [36].

One can use a LINAC in the entry in the Synchrotron and one at out (figure 98). To use a small entrance LINAC, between him and synchrotron, one put an additional speed circuit in a stadium form (fig. 98).

The end LINAC can be reduced if one put more end LINACs. See diagram below (fig. 98) [36].

Fig. 98. *A high energy synchrotron schema.*

Speaking about a new ionic engine means to speak about a new aircraft (fig 99) [36].

This ship has a circular particle accelerator (a synchrotron)

This ship has first a circular particle accelerator (a synchrotron), and at the end a big linear particle accelerator (a big LINAC)

This ship has two circular particle accelerators (two synchrotrons)

This ship has first a circular particle accelerator (a synchrotron), and at the end two big linear particle accelerators (two big LINAC)

Fig. 99. Some flying synchrotron prototypes.

This book (shortly) presents the actual ionic engines (called ion thrusters) and the new ionic (pulse) engines proposed by the authors.

Ionic engine (ion thruster, which accelerates the positive ions through a potential difference) is about 10 times more effective than classic system based on combustion [36].

We can still improve the efficiency of 10-50 times if one uses pulses of positive ions accelerated in a cyclotron mounted on the ship; the efficiency can easily grow for 1000 times if the positive ions will be accelerated in a high energy synchrotron, synchrocyclotron or isochronous cyclotron (1-100 GeV) [36].

Future (ionic) engine will have mandatory a circular particle accelerator (high or very high energy).

We can thus increase the speed and autonomy of the ship using a less quantity of fuel and power. One can use synchrotron radiation (synchrotron light, high intensity beams), like high intensity (X-ray or Gamma ray) radiation, as well. In this case will be a beam engine (not an ionic engine), it'll use only the power (energy, which can be solar energy, nuclear energy, or both) and so we will remove the fuel.

A linear particle accelerator (also called a LINAC) is an electrical device for the acceleration of subatomic particles. This sort of particle accelerator has many applications. It used recently as to an injector into a higher energy synchrotron at a dedicated experimental particle physics laboratory. In this, the big classic synchrotron is reduced to a ring surface (magnetic core).

The design of a LINAC depends on the type of particle that is being accelerated: electron, proton or ion.

It proposes using a powerful LINAC at the exit of synchrotron (especially when one accelerates electrons) to not lose energy by photons premature emission (figure 99).

One can use a LINAC in the entry in the Synchrotron and one at out (figure 98). To use a small entrance LINAC, between him and synchrotron, one put an additional speed circuit in a stadium form (fig. 98).

With a new ionic engine one builds a new aircraft, which can travel through water and.

This new aircraft will can accelerate directly, without an additional combustion engine and without gravity assists from other planets.

The compact engine, tightly closed, does not need air (oxygen).

The ship passes through clouds of dust without any problems.

The passage from the air into the water and vice versa can be peaceful or even accelerated, without any problems.

Ionic engine (ion thruster) has 2 major advantages (a) and 2 disadvantages (b) compared with chemical propulsion; (a) the impulse and energy per unit of fuel used are much higher; 1-the increased impulse generates a higher speed (velocity; so we can walk longer distances in a short time), 2-the high energy decreases fuel consumption and increase the autonomy of the ship; (b) generate force and acceleration are very small; we can not defeat any forces of resistance to lodging by atmosphere and we have no chance to exceed gravitational forces - ship will not leave a planet (or fall on it) using the ion thruster (It required an additional motor). Vacuum ship acceleration is possible but only with very small acceleration.

Increasing more the energy (and also the impulse) can reach the necessary forces and acceleration (the growth will need to be very high, 100 PeV-1000 PeV).

Particles energy increased can be made with accelerators circular and or modern linear. Particles energy increased will be huge and in addition will need to grow and the flow of accelerated particles (and the tor diameter; if one increases enough the flow, the necessary energy increased, will be less, 10 GeV-10 TeV).

We can increase the number of particles per pulse, the number of pulses per second, and the particle energy.

Immediate consequence of increasing particle energy will be the increasing of speeds and autonomy of the ship. Now we can achieve huge speeds in a very short time. The ship will pass through any atmosphere (including water) with great ease. The ship can take off or land directly.

Initially one can use to ships the old forms (the old design) which adapts and the accelerator(s).

This idea was presented by the author from 1996, but it has been published and copyrighted since 2009 [36].

Although its implementation stage is probably slightly more advanced, it could not be reached to far yet because of the limitations of the current accelerators.

Particle accelerator

A particle accelerator is a device that uses electromagnetic fields to propel charged particles to high speeds and to contain them in well-defined beams (fig. 100).

An ordinary CRT television set is a simple form of accelerator.

There are two basic types: electrostatic and oscillating field accelerators.

Fig. 100. *A 1960s single stage 2 MeV linear Van de Graaff accelerator, here opened for maintenance.*

In the early 20th century, cyclotrons were commonly referred to as atom smashers.

Despite the fact that modern colliders actually propel subatomic particles—atoms themselves now being relatively simple to disassemble without an accelerator—the term persists in popular usage when referring to particle accelerators in general.

Beams of high-energy particles are useful for both fundamental and applied research in the sciences (fig. 101), and also in many technical and industrial fields unrelated to fundamental research. It has been estimated that there are approximately 26,000 accelerators worldwide. Of these, only about 1% are research machines with energies above 1 GeV of the sort which are the main focus of this article. Of the rest, about 44% are for radiotherapy, 41% for ion implantation, 9% for industrial processing and research, and 4% for biomedical and other low-energy research.

Fig. 101. *Beamlines leading from the Van de Graaff accelerator to various experiments, in the basement of the Jussieu Campus in Paris.*

For the most basic inquiries into the dynamics and structure of matter, space, and time, physicists seek the simplest kinds of interactions at the highest possible energies. These typically entail particle energies of many GeV, and the interactions of the simplest kinds of particles: leptons (e.g. electrons and positrons) and quarks for the matter, or photons and gluons for the field quanta. Since isolated quarks are experimentally unavailable due to color confinement, the simplest available experiments involve the interactions of, first, leptons with each other, and second, of leptons with nucleons, which are composed of quarks and gluons. To study the collisions of quarks with each other, scientists resort to collisions of nucleons, which at high energy may be usefully considered as essentially 2-body interactions of the quarks and gluons of which they are composed. Thus elementary particle physicists tend to use machines creating beams of electrons, positrons, protons, and anti-protons, interacting with each other or with the simplest nuclei (e.g., hydrogen or deuterium) at the highest possible energies, generally hundreds of GeV or more.

Nuclear physicists and cosmologists may use beams of bare atomic nuclei, stripped of electrons, to investigate the structure, interactions, and properties of the nuclei themselves, and of condensed matter at extremely high temperatures and densities, such as might have occurred in the first moments of the Big Bang. These investigations often involve collisions of heavy nuclei – of atoms like iron or gold – at energies of several GeV per nucleon. At lower energies, beams of accelerated nuclei are also used in medicine, as for the treatment of cancer.

Besides being of fundamental interest, high energy electrons may be coaxed into emitting extremely bright and coherent beams of high energy photons – ultraviolet and X ray – via synchrotron radiation, which photons have numerous uses in the study of atomic structure, chemistry, condensed matter physics, biology, and technology.

Examples include the ESRF in Europe, which has recently been used to extract detailed 3-dimensional images of insects trapped in amber.

Thus there is a great demand for electron accelerators of moderate (GeV) energy and high intensity.

Low-energy machines

Everyday examples of particle accelerators are cathode ray tubes found in television sets and X-ray generators. These low-energy accelerators use a single pair of electrodes with a DC voltage of a few thousand volts between them. In an X-ray generator, the target itself is one of the electrodes. A low-energy particle accelerator called an ion implanter is used in the manufacture of integrated circuits.

High-energy machines

DC accelerator types capable of accelerating particles to speeds sufficient to cause nuclear reactions are Cockcroft-Walton generators or voltage multipliers (fig. 102), which convert AC to high voltage DC, or Van de Graaff generators that use static electricity carried by belts.

Fig. 102. In early particle accelerators a Cockcroft-Walton voltage multiplier was responsible for voltage multiplying. This piece of the accelerator helped in the development of the atomic bomb. Built in 1937 by Philips of Eindhoven it currently resides in the National Science Museum in London, England.

The largest and most powerful particle accelerators, such as the RHIC, the Large Hadron Collider (LHC) at CERN (which came on-line in mid-November 2009, fig. 103) and the Tevatron (fig. 104), are used for experimental particle physics.

Fig. 103. *The Large Hadron Collider (LHC) at CERN.*

The Tevatron is a circular particle accelerator in the United States, at the Fermi National Accelerator Laboratory, just east of Batavia, Illinois, and is the second highest energy particle collider in the world after the Large Hadron Collider (LHC, fig. 204). The Tevatron is a synchrotron that accelerates protons and antiprotons in a 6.28 km (3.90 miles) ring to energies of up to 1 TeV, hence the name. The Tevatron was completed in 1983 at a cost of $120 million ($265 million today) and has been regularly upgraded since then. (The 'Energy Doubler', as it was known then, produced its first accelerated beam — 512 GeV — on July 3, 1983.) The Main Injector was the most substantial addition, built over five years from 1994 at a cost of $290 million ($639 million today).

On January 10, 2011, it was announced that the Tevatron will cease operations at the end of September, 2011, as it has been made obsolete by the LHC, which began operations in early 2010. The main ring of the Tevatron will probably be reused in future experiments, and its components may be transferred to other particle accelerators.

Fig. 104. *The Tevatron (background) and Main Injector rings.*

Particle accelerators can also produce proton beams, which can produce proton-rich medical or research isotopes as opposed to the neutron-rich ones made in fission reactors; however, recent work has shown how to make 99Mo, usually made in reactors, by accelerating isotopes of hydrogen, although this method still requires a reactor to produce tritium. An example of this type of machine is LANSCE at Los Alamos (fig. 105).

Los Alamos Neutron Science Center (LANSCE) is the signature experimental science facility at Los Alamos National Laboratory.

Built more than 30 years ago, LANSCE was an innovative accelerator design conceived by Manhattan Project veteran scientist Louis Rosen to be a world-class neutron scattering facility. To complement the studies into the nature of elementary particles conducted at other institutions in the 1970s, the then-new facility would produce the highest-intensity proton beam in the world to explore the fundamental forces of nature at medium energies.

Fig. 105. *Los Alamos Linear Accelerator (LINAC).*

The diversity of research enabled by LANSCE is rivaled by few facilities anywhere in the world—LANSCE helps researchers solve problems in fields from biological science to materials science to fundamental physics.

A unique attribute of LANSCE is that its five major facilities (Lujan Neutron Scattering Center, Weapons Neutron Research Facility, Proton Radiography Facility, Isotope Production Facility, and Ultracold Neutron Source) can operate simultaneously.

On-going research at LANSCE includes medical applications, semiconductor development, nuclear deterrence, and energy security.

The core LANSCE facility (fig. 105) is one of the nation's most powerful proton linear accelerators or LINAC. The 800-mega-electron-volt (800 MeV) LINAC provides beam current, simultaneously, to five major facilities with unique capabilities: the Proton Radiography (pRad) facility that supports NNSA Defense Program (DP) missions; the Weapons Neutron Research facility (WNR) that supports DP missions; the Lujan Neutron Scattering Center that supports DP and DOE Office of Science (SC) missions; the Isotope Production Facility (IPF) that supports the Office of Nuclear Physics within the SC; and the Ultra-cold Neutrons (UCN) facility that supports SC missions.

The LINAC at LANSCE has served the nation since 1972, providing the beam current required by all the experimental areas that support NNSA-DP and other DOE missions. The LINAC's capability to reliably deliver beam current is the key to the LANSCE's ability to do research–and thus the key to meeting NNSA and DOE mission deliverables.

Electrostatic particle accelerators

Historically, the first accelerators used simple technology of a single static high voltage to accelerate charged particles. While this method is still extremely popular today, with the electrostatic accelerators greatly out-numbering any other type, they are more suited to lower energy studies owing to the practical voltage limit of about 30 MV (when the accelerator is placed in a gas with high dielectric strength, such as sulfur hexafluoride, allowing the high voltage). The same high voltage can be used twice in a tandem accelerator if the charge of the particles can be reversed while they are inside the terminal; this is possible with the acceleration of atomic nuclei by first adding an extra electron or forming an anionic (negatively charged) chemical compound, and then putting the beam through a thin foil to strip off electrons inside the high voltage conducting terminal, making a beam of positive charge.

Although electrostatic accelerators accelerate particles along a straight line, the term linear accelerator is more often associated with accelerators that use oscillating rather than static electric fields. Thus, many accelerators arranged in a straight line are not termed "linear accelerators" but rather "electrostatic accelerators" to differentiate the two cases.

Oscillating field particle accelerators

Due to the high voltage ceiling imposed by electrical discharge, in order to accelerate particles to higher energies, techniques involving more than one lower, but oscillating, high voltage sources are used. The electrodes can either be arranged to accelerate particles in a line or circle, depending on whether the particles are subject to a magnetic field while they are accelerated, causing their trajectories to arc.

Linear particle accelerators

In a linear accelerator (linac), particles are accelerated in a straight line with a target of interest at one end. They are often used to provide an initial low-energy kick to particles before they are injected into circular accelerators.

The longest linac in the world is the Stanford Linear Accelerator, SLAC, which is 3 km (2 miles) long. SLAC is an electron-positron collider.

Linear high-energy accelerators use a linear array of plates (or drift tubes) to which an alternating high-energy field is applied. As the particles approach a plate they are accelerated towards it by an opposite polarity charge applied to the plate.

As they pass through a hole in the plate, the polarity is switched so that the plate now repels them and they are now accelerated by it towards the next plate.

Normally a stream of "bunches" of particles are accelerated, so a carefully controlled AC voltage is applied to each plate to continuously repeat this process for each bunch.

As the particles approach the speed of light the switching rate of the electric fields becomes so high that they operate at microwave frequencies, and so RF cavity resonators are used in higher energy machines instead of simple plates.

Linear accelerators are also widely used in medicine, for radiotherapy and radiosurgery. Medical grade LINACs accelerate electrons using a klystron and a complex bending magnet arrangement which produces a beam of 6-30 million electron-volt (MeV) energy.

The electrons can be used directly or they can be collided with a target to produce a beam of X-rays. The reliability, flexibility and accuracy of the radiation beam produced has largely supplanted the older use of Cobalt-60 therapy as a treatment tool.

Circular or cyclic accelerators

In the circular accelerator, particles move in a circle until they reach sufficient energy. The particle track is typically bent into a circle using electromagnets.

The advantage of circular accelerators over linear accelerators (linacs) is that the ring topology allows continuous acceleration, as the particle can transit indefinitely.

Another advantage is that a circular accelerator is smaller than a linear accelerator of comparable power (i.e. a linac would have to be extremely long to have the equivalent power of a circular accelerator).

Depending on the energy and the particle being accelerated, circular accelerators suffer a disadvantage in that the particles emit synchrotron radiation.

When any charged particle is accelerated, it emits electromagnetic radiation and secondary emissions. As a particle traveling in a circle is always accelerating towards the center of the circle, it continuously radiates towards the tangent of the circle. This radiation is called synchrotron light and depends highly on the mass of the accelerating particle. For this reason, many high energy electron accelerators are linacs. Certain accelerators (synchrotrons) are however built specially for producing synchrotron light (X-rays).

Since the special theory of relativity requires that matter always travels slower than the speed of light in a vacuum, in high-energy accelerators, as the energy increases the particle speed approaches the speed of light as a limit, but never attains it.

Therefore particle physicists do not generally think in terms of speed, but rather in terms of a particle's energy or momentum, usually measured in electron volts (eV).

An important principle for circular accelerators, and particle beams in general, is that the curvature of the particle trajectory is proportional to the particle charge and to the magnetic field, but inversely proportional to the (typically relativistic) momentum.

Cyclotrons

The earliest circular accelerators were cyclotrons, invented in 1929 by Ernest O. Lawrence (fig. 106) at the University of California, Berkeley.

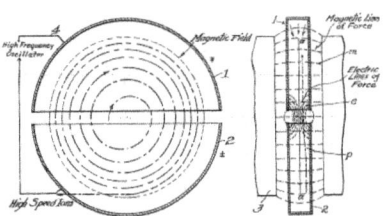

Fig. 106. *Diagram of cyclotron operation from Lawrence's 1934 patent.*

Cyclotrons have a single pair of hollow 'D'-shaped plates to accelerate the particles and a single large dipole magnet to bend their path into a circular orbit (fig. 107).

It is a characteristic property of charged particles in a uniform and constant magnetic field B that they orbit with a constant period, at a frequency called the cyclotron frequency, so long as their speed is small compared to the speed of light c.

This means that the accelerating D's of a cyclotron can be driven at a constant frequency by a radio frequency (RF) accelerating power source, as the beam spirals outwards continuously.

The particles are injected in the centre of the magnet and are extracted at the outer edge at their maximum energy (fig. 108).

Fig. 107. Beam of electrons moving in a circle. Lighting is caused by excitation of gas atoms in a bulb.

Cyclotrons reach an energy limit because of relativistic effects whereby the particles effectively become more massive, so that their cyclotron frequency drops out of synch with the accelerating RF.

Therefore simple cyclotrons can accelerate protons only to an energy of around 15 million electron volts (15 MeV, corresponding to a speed of roughly 10% of c), because the protons get out of phase with the driving electric field.

If accelerated further, the beam would continue to spiral outward to a larger radius but the particles would no longer gain enough speed to complete the larger circle in step with the accelerating RF. Cyclotrons are nevertheless still useful for lower energy applications.

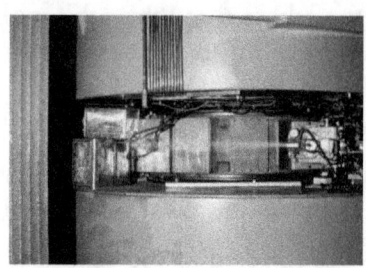

***Fig. 108.** 60-inch cyclotron, circa 1939, showing a beam of accelerated ions (likely protons or deuterons) escaping the accelerator and ionizing the surrounding air causing a blue glow.*

Synchrocyclotrons and isochronous cyclotrons

There are ways of modifying the classic cyclotron to increase the energy limit.

This may be done in a continuous beam, constant frequency, machine by shaping the magnet poles so to increase magnetic field with radius.

Then higher energy particles travel a shorter distance in each orbit than they otherwise would, and can remain in phase with the accelerating field. Such machines are called isochronous cyclotrons. Their advantage is that they can deliver continuous beams of higher average intensity, which is useful for some applications.

The main disadvantages are the size and cost of the large magnet needed, and the difficulty in achieving the higher field required at the outer edge. Another possibility, the synchrocyclotron (fig. 109), accelerates the particles in bunches, in a constant B field, but reduces the RF accelerating field's frequency so as to keep the particles in step as they spiral outward.

This approach suffers from low average beam intensity due to the bunching, and again from the need for a huge magnet of large radius and constant field over the larger orbit demanded by high energy.

Fig. 109. *A magnet in the synchrocyclotron at the Orsay proton therapy center.*

Betatrons

Another type of circular accelerator, invented in 1940 for accelerating electrons, is the Betatron, a concept which originates ultimately from Norwegian-German scientist Rolf Widerøe.

These machines, like synchrotrons, use a donut-shaped ring magnet (see below) with a cyclically increasing B field, but accelerate the particles by induction from the increasing magnetic field, as if they were the secondary winding in a transformer, due to the changing magnetic flux through the orbit.

Achieving constant orbital radius while supplying the proper accelerating electric field requires that the magnetic flux linking the orbit be somewhat independent of the magnetic field on the orbit, bending the particles into a constant radius curve.

These machines have in practice been limited by the large radiative losses suffered by the electrons moving at nearly the speed of light in a relatively small radius orbit (fig. 110).

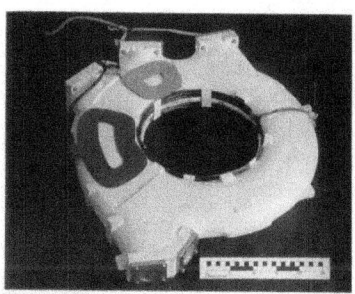

Fig. 110. *A Betatron.*

Synchrotrons

To reach still higher energies, with relativistic mass approaching or exceeding the rest mass of the particles (for protons, billions of electron volts or GeV), it is necessary to use a synchrotron (fig. 111). This is an accelerator in which the particles are accelerated in a ring of constant radius. An immediate advantage over cyclotrons is that the magnetic field need only be present over the actual region of the particle orbits, which is very much narrower than the diameter of the ring. (The largest cyclotron built in the US had a 184-inch-diameter (4.7 m) magnet pole, whereas the diameter of the LEP and LHC is nearly 10 km. The aperture of the two beams of the LHC is of the order of a millimeter.)

However, since the particle momentum increases during acceleration, it is necessary to turn up the magnetic field B in proportion to maintain constant curvature of the orbit.

In consequence synchrotrons cannot accelerate particles continuously, as cyclotrons can, but must operate cyclically, supplying particles in bunches, which are delivered to a target or an external beam in beam "spills" typically every few seconds.

Since high energy synchrotrons do most of their work on particles that are already traveling at nearly the speed of light c, the time to complete one orbit of the ring is nearly constant, as is the frequency of the RF cavity resonators used to drive the acceleration.

Note also a further point about modern synchrotrons: because the beam aperture is small and the magnetic field does not cover the entire area of the particle orbit as it does for a cyclotron, several necessary functions can be separated.

Instead of one huge magnet, one has a line of hundreds of bending magnets, enclosing (or enclosed by) vacuum connecting pipes. The design of synchrotrons was revolutionized in the early 1950s with the discovery of the strong focusing concept.

The focusing of the beam is handled independently by specialized quadrupole magnets, while the acceleration itself is accomplished in separate RF sections, rather similar to short linear accelerators. Also, there is no necessity that cyclic machines be circular, but rather the beam pipe may have straight sections between magnets where beams may collide, be cooled, etc. This has developed into an entire separate subject, called "beam physics" or "beam optics". More complex modern synchrotrons such as the Tevatron, LEP, and LHC may deliver the particle bunches into storage rings of magnets with constant B, where they can continue to orbit for long periods for experimentation or further acceleration.

The highest-energy machines such as the Tevatron and LHC are actually accelerator complexes, with a cascade of specialized elements in series, including linear accelerators for initial beam creation, one or more low energy synchrotrons to reach intermediate energy, storage rings where beams can be accumulated or "cooled" (reducing the magnet aperture required and permitting tighter focusing; see beam cooling), and a last large ring for final acceleration and experimentation.

Fig. 111. *The interior of the Australian Synchrotron facility. Dominating the image is the storage ring, showing the optical diagnostic beamline at front right. In the middle of the storage ring is the booster synchrotron and linac.*

Particles are injected into the main ring at substantial energies by either a linear accelerator or by an intermediate synchrotron which is in turn fed by a linear accelerator.

The "linac" is in turn fed by particles accelerated to intermediate energy by a simple high voltage power supply, typically a Cockcroft-Walton generator.

Starting from an appropriate initial value determined by the injection velocity the magnetic field is then increased. The particles pass through an electrostatic accelerator driven by a high alternating voltage.

At particle speeds not close to the speed of light the frequency of the accelerating voltage can be made roughly proportional to the current in the bending magnets.

A finer control of the frequency is performed by a servo loop which responds to the detection of the passing of the traveling group of particles.

At particle speeds approaching light speed the frequency becomes more nearly constant, while the current in the bending magnets continues to increase.

The maximum energy that can be applied to the particles (for a given ring size and magnet count) is determined by the saturation of the cores of the bending magnets (the point at which increasing current does not produce additional magnetic field).

One way to obtain additional power is to make the torus larger and add additional bending magnets.

This allows the amount of particle redirection at saturation to be less and so the particles can be more energetic. Another means of obtaining higher power is to use superconducting magnets, these not being limited by core saturation.

Large synchrotrons

One of the early large synchrotrons, now retired, is the Bevatron, constructed in 1950 at the Lawrence Berkeley Laboratory.

The name of this proton accelerator comes from its power, in the range of 6.3 GeV (then called BeV for billion electron volts; the name predates the adoption of the SI prefix giga).

A number of heavy elements, unseen in the natural world, were first created with this machine.

This site is also the location of one of the first large bubble chambers used to examine the results of the atomic collisions produced here.

Another early large synchrotron is the Cosmotron built at Brookhaven National Laboratory which reached 3.3 GeV in 1953.

Until August 2008, the highest energy synchrotron in the world was the Tevatron, at the Fermi National Accelerator Laboratory, in the United States.

It accelerates protons and antiprotons to slightly less than 1 TeV of kinetic energy and collides them together (fig. 104).

The Large Hadron Collider (LHC, fig. 103), which has been built at the European Laboratory for High Energy Physics (CERN), has roughly seven times this energy (so proton-proton collisions occur at roughly 14 TeV).

It is housed in the 27 km tunnel which formerly housed the Large Electron Positron (LEP) collider, so it will maintain the claim as the largest scientific device ever built.

The LHC will also accelerate heavy ions (such as lead) up to an energy of 1.15 PeV.

The largest device of this type seriously proposed was the Superconducting Super Collider (SSC), which was to be built in the United States.

This design, like others, used superconducting magnets which allow more intense magnetic fields to be created without the limitations of core saturation.

While construction was begun, the project was cancelled in 1994, citing excessive budget overruns — this was due to naïve cost estimation and economic management issues rather than any basic engineering flaws.

It can also be argued that the end of the Cold War resulted in a change of scientific funding priorities that contributed to its ultimate cancellation.

While there is still potential for yet more powerful proton and heavy particle cyclic accelerators, it appears that the next step up in electron beam energy must avoid losses due to synchrotron radiation. This will require a return to the linear accelerator, but with devices significantly longer than those currently in use.

There is at present a major effort to design and build the International Linear Collider (ILC), which will consist of two opposing linear accelerators, one for electrons and one for positrons.

These will collide at a total center of mass energy of 0.5 TeV.

However, synchrotron radiation also has a wide range of applications (see synchrotron light) and many 2nd and 3rd generation synchrotrons have been built especially to harness it.

The largest of those 3rd generation synchrotron light sources are the European Synchrotron Radiation Facility (ESRF) in Grenoble, France, the Advanced Photon Source (APS) near Chicago, USA, and

SPring-8 in Japan, accelerating electrons up to 6, 7 and 8 GeV, respectively.

Synchrotrons which are useful for cutting edge research are large machines, costing tens or hundreds of millions of dollars to construct, and each beamline (there may be 20 to 50 at a large synchrotron) costs another two or three million dollars on average.

These installations are mostly built by the science funding agencies of governments of developed countries, or by collaborations between several countries in a region, and operated as infrastructure facilities available to scientists from universities and research organisations throughout the country, region, or world.

More compact models, however, have been developed, such as the Compact Light Source.

SOLEIL ("Sun" in French) is a synchrotron facility near Paris, France. It performed its first acceleration of electrons on May 14, 2006.

The name SOLEIL is a backronym for Source optimisée de lumière d'énergie intermédiaire du LURE (LURE optimised intermediary energy light source), LURE meaning Laboratoire pour l'utilisation du rayonnement électromagnétique.

The facility is run by a civil corporation held by the French National Centre for Scientific Research (CNRS) and the Commissariat à l'Énergie Atomique (CEA), two French national research agencies.

It is located in Saint-Aubin in the Essonne département, a south-western suburb of Paris, near Gif-sur-Yvette and Saclay, which host other facilities for nuclear and particle physics (fig. 112).

Fig. 112. *Modern industrial-scale synchrotrons can be very large (here, Soleil near Paris).*

The Advanced Photon Source (APS) at Argonne National Laboratory (in Argonne, IL, USA, fig. 113) is a national synchrotron-radiation light source research facility funded by the United States Department of Energy Office of Science. Argonne National Laboratory is managed by UChicago Argonne LLC, which is composed of the University of Chicago and Jacobs Engineering Group, Inc.

Fig. 113. The Advanced Photon Source (APS) at Argonne National Laboratory (in Argonne, IL, USA) is a national synchrotron-radiation light source research facility funded by the United States Department of Energy Office of Science.

Using high-brilliance X-ray beams from the APS, members of the international synchrotron-radiation research community conduct forefront basic and applied research in the fields of materials science and biological science; physics and chemistry; environmental, geophysical and planetary science; and innovative X-ray instrumentation.

ISIS is a pulsed neutron and muon source. It is situated at the Rutherford Appleton Laboratory on the Harwell Science and Innovation Campus in Oxfordshire, United Kingdom and is part of the Science and Technology Facilities Council (fig. 114). It uses the techniques muon spectroscopy and neutron scattering to probe the structure and dynamics of condensed matter on a microscopic scale ranging from the subatomic to the macromolecular.

Hundreds of experiments are performed annually at ISIS by visiting researchers from around the world, in diverse science areas including physics, chemistry, materials engineering, earth sciences, biology and archaeology.

Fig. 114. *Another view of the ISIS experimental hall for Target Station 1.*

Neutrons are uncharged constituents of atoms and penetrate materials well, deflecting only from the nuclei of atoms.

The statistical accumulation of deflected neutrons at different positions beyond the sample can be used to find the structure of a material, and the loss or gain of energy by neutrons can reveal the dynamic behaviour of parts of a sample, for example diffusive processes in solids.

At ISIS the neutrons are created by accelerating 'bunches' of protons in a synchrotron, then colliding these with a heavy tungsten metal target, under a constant cooling load to dissipate the heat from the 160 kW proton beam.

The impacts cause neutrons to spall off the tungsten atoms, and the neutrons are channelled through guides, or beamlines, to about 20 instruments, individually optimised for the study of different types of matter.

The target station and most of the instruments are set in a large hall.

Neutrons are a dangerous form of radiation, so the target and beamlines are heavily shielded with concrete.

ISIS produces muons by colliding a fraction of the proton beam with a graphite target, producing pions which decay rapidly into muons, delivered in a spin-polarised beam to sample stations.

Fig. 115. *Laboratório Nacional de Luz Síncrotron (LNLS) is the Brazilian National Laboratory of Synchrotron Light.*

Laboratório Nacional de Luz Síncrotron (LNLS) is the Brazilian National Laboratory of Synchrotron Light (fig. 115), a research institution on physics, chemistry, material science and life sciences. It is located in the Campinas' sub-district of Barão Geraldo, state of São Paulo, Brazil. The Center, which is operated by the Brazilian Association for Synchrotron Light Technology (ABTLuS) under a contract with the National Research Council (CNPq) and the Ministry of Science and Technology of Brazil, has the only particle accelerator (a synchrotron) in Latin America, which was designed and built in Brazil by a team of physicists, technicians and engineers.

The Bevatron was a historic particle accelerator — specifically, a weak-focusing proton synchrotron — at Lawrence Berkeley National Laboratory which began operating in 1954 (fig. 116).

The antiproton was discovered there in 1955, resulting in the 1959 Nobel Prize in physics for Emilio Segrè and Owen Chamberlain.

It accelerated protons into a fixed target, and was named for its ability to impart energies of billions of eV. (Billions of eV Synchrotron.)

Fig. 116. *Edwin McMillan and Edward Lofgren on the shielding of the Bevatron. The shielding was only added later, after initial operations.*

At the time the Bevatron was designed, it was strongly suspected but not known, that each particle had a corresponding anti-particle of opposite charge, identical in all other respects, a property known as charge symmetry. The anti-electron, or positron had been first observed in the early 1930s, and theoretically understood as a consequence of the Dirac equation at about the same time. Following World War II, positive and negative muons and pions were observed in cosmic-ray interactions seen in cloud chambers and stacks of nuclear photographic emulsions. The Bevatron was built to be energetic enough to create antiprotons, and thus test the hypothesis that every particle has a corresponding anti-particle. The antineutron was discovered soon thereafter by Oreste Piccioni and co-workers, also at the Bevatron. Confirmation of the charge symmetry conjecture in 1955 led to the Nobel Prize for physics being awarded to Emilio Segrè and Owen Chamberlain in 1960.

Shortly after the Bevatron came into use, it was recognized . that parity was not conserved in the weak interactions, which led to resolution of the tau-theta puzzle, the understanding of strangeness, and the establishment of CPT symmetry as a basic feature of relativistic quantum field theories.

In order to create antiprotons (mass ~938 MeV/c2) in collisions with nucleons in a stationary target while conserving both energy and momentum, a proton beam energy of approximately 6.2 GeV is required. At the time it was built, there was no known way to confine a particle beam to a narrow aperture, so the beam space was about four square feet in cross section. The combination of beam aperture and energy required a huge, 10,000 ton iron magnet, and a very large vacuum system.

A large motor/generator system was used to ramp up the magnetic field for each cycle of acceleration. At the end of each cycle, after the beam was used or extracted, the large magnetic field energy was returned to spin up the motor, which was then used as a generator to power the next cycle, conserving energy; the entire process required about five seconds. The characteristic rising and falling, wailing, sound of the motor-generator system could be heard in the entire complex when the machine was in operation.

In the years following the antiproton discovery, much pioneering work was done here using beams of protons extracted from the accelerator proper, to hit targets and generate secondary beams of elementary particles, not only protons but also neutrons, pions, "strange particles", and many others.

The extracted particle beams, both the primary protons and secondaries, could in turn be passed for further study through various targets and specialized detectors, notably the liquid hydrogen bubble chamber. Many thousands of particle interactions, or "events", were photographed, measured, and studied in detail with an automated system of large measuring machines (known as "Frankensteins") allowing human operators (typically the wives of graduate students) to mark points along the particle tracks and punch their coordinates into IBM cards, using a foot pedal. The cards decks were then analyzed by early-generation computers, which reconstructed the three-dimensional tracks through the magnetic fields, and computed the momenta and energy of the particles. Computer programs, extremely complex for their time, then fitted the track data associated with a given event to estimate the energies, masses, and identities of the particles produced (fig. 117).

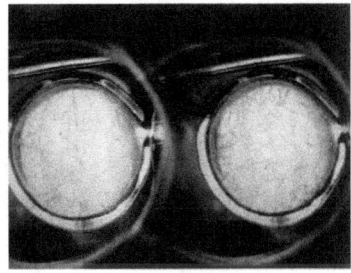

Fig. 117. First tracks observed in liquid hydrogen bubble chamber at the Bevatron.

This period, when hundreds of new particles and excited states were suddenly revealed, marked the beginning of a new era in elementary particle physics. Luis Alvarez inspired and directed much of this work, for which he received the Nobel Prize in physics in 1968.

The **Bevatron** received a new lease on life in 1971, when it was joined to the SuperHILAC linear accelerator as an injector for heavy ions. The combination was conceived by Albert Ghiorso, who named it the **Bevalac**. It could accelerate any nuclei in the periodic table to relativistic energies. It was finally decommissioned in 1993.

The next generation of accelerators used "strong focusing", and required much smaller apertures, and thus much cheaper magnets. The AGS (Alternating Gradient Synchrotron) at Brookhaven was the first next-generation machine, with an aperture roughly an order of magnitude less in both transverse directions, and reaching 30 GeV proton energy, yet with a less massive magnet ring. For comparison, the circulating beams in the Large Hadron Collider, the latest and greatest descendant of the Bevatron (with ~11,000 times higher energy and enormously higher intensity), are confined to a space on the order of 1 mm in cross-section, and focused down to 16 micrometres at the intersection collision regions, while the field of the bending magnets is only about five times higher.

The demolition begun in 2009 by Clauss Construction of Lakeside CA and is scheduled for completion in 2011.

The Advanced Light Source (ALS) at Lawrence Berkeley National Laboratory in Berkeley, California is a synchrotron light source. Built from 1987 to 1993, it currently employs 210 scientists and staff. Part of the building in which it is housed was completed in 1942 for a 4.67 m (184 in) cyclotron, designed by Arthur Brown, Jr. (designer of the Coit Tower in San Francisco) and built by Ernest O. Lawrence. Today, the expanded building houses the ALS, a U.S. Department of Energy national user facility that attracts scientists from around the world.

The ALS is a national user facility that generates intense light for scientific and technological research. As one of the world's brightest sources of ultraviolet and soft x-ray beams--and the world's first third-generation synchrotron light source in its energy range--the ALS makes what were thought to be impossible studies possible. The facility welcomes over 2000 researchers every year from universities, industries, and government laboratories around the world. It is funded by the U.S. Department of Energy's Office of Basic Energy Sciences.

The ALS has over forty beamlines which simultaneously perform a wide range of science. Any qualified scientist can propose to use the ALS beamlines.

Proposals are peer-reviewed and top-ranked proposals are allocated beam time. The ALS does not charge for beam time if the user's research is nonproprietary (results are published in open literature).

The Cosmotron was a particle accelerator, specifically a proton synchrotron, at Brookhaven National Laboratory.

Its construction was approved by the U.S. Atomic Energy Commission in 1948, it reached its full energy in 1953, and it continued running until 1968.

It was the first particle accelerator to impart kinetic energy in the range of GeV to a single particle, accelerating protons to 3.3 GeV.

It was also the first accelerator to allow the extraction of the particle beam for experiments located physically outside the accelerator.

It was used to observe a number of mesons previously seen only in cosmic rays, and to make the first discoveries of heavy, unstable particles (called V particles at the time).

The National Synchrotron Light Source (NSLS) at Brookhaven National Laboratory (BNL) in Upton, New York is a national user research facility funded by the U.S. Department of Energy (DOE). The NSLS, is considered a second generation synchrotron, and was built beginning in 1978 and finished in 1984 (fig. 118).

The NSLS experimental floor consists of two electron storage rings: an X-Ray ring and a VUV (Vacuum Ultra Violet) Ring which provide intense focused light spanning the electromagnetic spectrum from the infrared through x-rays. The properties of this light and the specially designed experimental stations, called beamlines, allow scientists in many fields of research to perform experiments not otherwise possible at their own laboratories.

Fig. 118. *Aerial view of the NSLS.*

The Stanford Synchrotron Radiation Lightsource (formerly Stanford Synchrotron Radiation Laboratory), a division of SLAC National Accelerator Laboratory, is operated by Stanford University for the Department of Energy. SSRL is a National User Facility which provides synchrotron radiation, a name given to electromagnetic radiation in the x-ray, ultraviolet, visible and infrared realms produced by electrons circulating in a storage ring (SPEAR) at nearly the speed of light. These extremely bright x-rays can be used to investigate various forms of matter ranging from objects of atomic and molecular size to man-made materials with unusual properties. The obtained information and knowledge is of great value to society, with impact in areas such as the environment, future technologies, health, and education. The SSRL provides experimental facilities to some 2,000 academic and industrial scientists working in such varied fields as drug design, environmental cleanup, electronics, and x-ray imaging. It is located in southern San Mateo County, just outside the city of Menlo Park.

In 1972 the first x-ray beamline was constructed by Ingolf Lindau and Piero Pianetta as literally a "hole in the wall" extending off of the SPEAR (Stanford Positron Electron Asymmetric Ring) storage ring. At that time, the SPEAR had been built in an era of particle colliders, where physicists were more interested in smashing particles together in hope of discovering antimatter then in using x-ray radiation for solid state physics and chemistry. From those meager beginnings the Stanford Synchrotron Radiation Project (SSRP) began. Within a short time SSRP had five experimental hutches sharing the radiation of only a few inches of the curved SPEAR dipole magnets. Each one of those stations was equipped with a monochromator to select the radiation of interest, and experimenters would bring their samples and end stations from all over the world to study the unique effects only achieved through synchrotron radiation. Today the SPEAR storage ring is dedicated completely to the Stanford Synchrotron Radiation Lightsource as part of the SLAC National Accelerator Laboratory facility. SSRL currently operates 24/7 for about nine solid months out of the year; the remaining time is used for major maintenance and upgrades where direct access to the storage ring is needed. There are over 30 unique experimental stations which are made available to users from universities, government labs, and industry from all over the world.

The Proton Synchrotron (PS) is the first major particle accelerator at CERN, built as a 28 GeV proton accelerator in the late 1950s and put into operation in 1959. It takes the protons from the Proton Synchrotron Booster at a kinetic energy of 1.4 GeV and lead ions from the Low Energy Ion Ring (LEIR) at 72 MeV per nucleon. It has been operated as an injector for the Intersecting Storage Rings

(ISR), the Super Proton Synchrotron (SPS) and the Large Electron-Positron Collider (LEP). Starting in November 2009, the PS machine delivers protons and will provide lead ion beams for the Large Hadron Collider (LHC). It has also been used as a particle source for other experiments, such as the Gargamelle bubble chamber for which it supplied a neutrino beam. This led to the discovery of the weak neutral current in 1974.

The PS machine is a circular accelerator with a circumference of 628.3 m. It is a versatile machine having accelerated protons, antiprotons, electrons, positrons and species of ions. Major upgrades have improved its performance by more than a factor of 1000 since 1959. The only main components remaining from its original installation some 50 years ago are the bending magnets and the buildings.

The Swiss Light Source is a synchrotron located at the Paul Scherrer Institute in Switzerland for producing electromagnetic radiation of high brightness (fig. 119). Planning started in 1991, project was approved in 1997, and first light from the storage ring was seen at December 15, 2000. The experimental program started in June 2001 and it is used for research in materials science, biology and chemistry.

Fig. 119. *Panorama of the SLS.*

The Synchrotron Radiation Source (SRS) at the Daresbury Laboratory in Cheshire, England was the first second-generation synchrotron radiation source to produce X-rays. The research facility provided synchrotron radiation to a large number (at one point 38) experimental stations and had an operating cost of approximately £20 million per annum.

SRS had been operated by the Science and Technology Facilities Council. The SRS was closed in August 2008 after 28 years of operation, and is presently being decommissioned, with no major plans for the re-use of the existing building.

Diamond Light Source is a synchrotron facility located in Oxfordshire, United Kingdom (fig. 120). *Its purpose is to produce intense beams of light whose special characteristics are useful in many areas of scientific research.* In particular it can be used to investigate the structure and properties of a wide range of materials (proteins, for example).

Diamond generates synchrotron light at wavelengths ranging from X-rays to the far infrared. This is also known as synchrotron radiation and is the electromagnetic radiation emitted by charged particles travelling near the speed of light. It is used in a huge variety of experiments to study the structure and behaviour of many different types of matter.

The particles Diamond uses are electrons travelling at an energy of 3 GeV round a 561.6 m circumference storage ring. The ring is not circular, but is shaped as a twenty-four-sided polygon. As the electrons pass through specially designed magnets at each vertex, their sudden change of direction causes them to emit an exceptionally bright beam of radiation. This is the synchrotron light used for experiments.

The electrons reach this high energy via a series of pre-accelerator stages before being injected into the storage ring:

an electron gun

a 100 MeV linear accelerator

a 100 MeV–3GeV booster synchrotron.

Diamond is housed in a silver toroidal building of 738m in circumference, covering an area in excess of 43,300 square metres, or the area of over six international Association football pitches. This contains the storage ring and a number of beamlines, with the linear accelerator and booster synchrotron housed in the centre of the ring. These are experimental stations where the synchrotron light's interaction with matter is used for research purposes. Seven were available when the Diamond became operational in 2007, with plans to build another fifteen between then and 2011 at an additional cost of £120 million. As of December 2010 there were eighteen in operation and more under construction.

The seven available when Diamond opened were:

extreme conditions beamline for studying materials under intense temperatures and pressures.

materials and magnetism beamline, set up to probe electronic and magnetic materials at the atomic level.

three macromolecular crystallography beamlines, for decoding the structure of complex biological samples, such as proteins.

microfocus spectroscopy beamline, able to map the chemical make up of complex materials such as moon rocks and geological samples.

nanoscience beamline, capable of imaging structures and devices at the scale of a few nanometres (millionths of a millimetre).

Diamond is intended ultimately to host up to forty beamlines, supporting the life, physical and environmental sciences.

On 13 September 2007, scientists from Cardiff University, led by Professor Tim Wess, found that the Diamond synchrotron could be used to discover hidden content of ancient documents by illumination without opening them (penetrating layers of parchment).

Fig. 120. Aerial view of The Diamond Synchrotron.

!A Synchrotron like this, it the size of a stadium (fig. 220), built like a flying ship, could carry from 30,000 to 50,000 passengers.!

Storage rings

For some applications, it is useful to store beams of high energy particles for some time (with modern high vacuum technology, up to many hours) without further acceleration.

This is especially true for colliding beam accelerators, in which two beams moving in opposite directions are made to collide with each other, with a large gain in effective collision energy.

Because relatively few collisions occur at each pass through the intersection point of the two beams, it is customary to first accelerate the beams to the desired energy, and then store them in storage rings (fig. 111), which are essentially synchrotron rings of magnets, with no significant RF power for acceleration.

A storage ring is a type of circular particle accelerator in which a continuous or pulsed particle beam may be kept circulating for a long period of time, up to many hours. Storage of a particular particle depends upon the mass, energy and usually charge of the particle being stored. Most commonly, storage rings are to store electrons, positrons, or protons.

The most common application of storage rings is to store electrons which then radiate synchrotron radiation. There are over 50 facilities based on electron storage rings in the world today, used for a variety of studies in chemistry and biology.

Storage rings are used to produce polarized high-energy electron beams through the Sokolov-Ternov effect. Arguably the best known application of storage rings is their use in particle accelerators and in particle colliders, in which two counter-rotating beams of stored particles are brought into collision at discrete locations, the results of the subatomic interactions being studied in a surrounding particle detector.

Examples of such facilities are LHC, LEP, PEP-II, KEKB, RHIC, Tevatron and HERA.

Technically speaking, a storage ring is a type of synchrotron. However, a conventional synchrotron serves to accelerate particles from a low to a high energy with the aid of radio-frequency accelerating cavities; a storage ring, as the name suggests, keeps particles stored at a constant energy, and radio-frequency cavities are only used to replace energy lost through synchrotron radiation and other processes.

Gerard K. O'Neill proposed the use of storage rings as building blocks for a collider in 1956. A key benefit of storage rings in this context is that the storage ring can accumulate a high beam flux from an injection accelerator that achieves a much lower flux.

A force must be applied to particles in such a way that they are constrained to move approximately in a circular path.

This may be accomplished using either dipole electrostatic or dipole magnetic fields, but because most storage rings store relativistic charged particles it turns out that it is most practical to utilise magnetic fields produced by dipole magnets (fig. 121).

However, electrostatic accelerators have been built to store very low energy particles, and quadrupole fields may be used to store (uncharged) neutrons; these are comparatively rare, however.

Fig. 121. *Different types of magnets used in the storage ring of the Australian Synchrotron. The larger yellow (white) one is a dipole magnet used to bend the electron beam and produce the synchrotron radiation. The green one is a sextupole magnet, the red one (at back) a quadrupole magnet - these are used to focus and steer the electron beam.*

Dipole magnets alone only provide what is called weak focusing, and a storage ring composed of only these sorts of magnetic elements results in the particles have a relatively large beam size.

Interleaving dipole magnets with an appropriate arrangement of quadrupole and sextupole magnets can give a suitable strong focusing system that can give a much smaller beam size.

The FODO and Chasman-Green lattice structures are simple examples of strong focusing systems, but there are many others.

Dipole and quadrupole magnets deflect different particle energies by differing amounts, a property called chromaticity by analogy with physical optics.

The spread of energies that is inherently present in any practical stored particle beam will therefore give rise to a spread of transverse and longitudinal focusing, as well as contributing to various particle beam instabilities.

Sextupole magnets (and higher order magnets) are used to correct for this phenomenon, but this in turn gives rise to nonlinear motion that is one of the main problems facing designers of storage rings.

As the bunches will travel many millions of kilometers (considering that they will be moving at near the speed of light for many hours), any residual gas in the beam pipe will result in many, many collisions.

This will have the effect of increasing the size of the bunch, and increasing the energy spread. Therefore, a better vacuum yields better beam dynamics.

Also, single large-angle scattering events from either the residual gas, or from other particles in the bunch (Touschek scattering), can eject particles far enough that they are lost on the walls of the accelerator vacuum vessel.

This gradual loss of particles is called beam lifetime, and means that storage rings must be periodically injected with a new complement of particles.

Injection of particles into a storage ring may be accomplished in a number of ways, depending on the application of the storage ring.

The simplest method uses one or more pulsed deflecting dipole magnets (injection kicker magnets) to steer an incoming train of particles onto the stored beam path; the kicker magnets are turned off before the stored train returns to the injection point, thus resulting in a stored beam.

This method is sometimes called single-turn injection.

Multi-turn injection allows accumulation of many incoming trains of particles, for example if a large stored current is required. For particles such as protons where there is no significant beam damping, each injected pulse is placed onto a particular point in the stored beam transverse or longitudinal phase space, taking care not to eject previously-injected trains by using a careful arrangement of beam deflection and coherent oscillations in the stored beam.

If there is significant beam damping, for example radiation damping of electrons due to synchrotron radiation, then an injected pulse may be placed on the edge of phase space and then left to damp in transverse phase space into the stored beam before injecting a further pulse. Typical damping times from synchrotron radiation are tens of milliseconds, allowing many pulses per second to be accumulated.

If extraction of particles is required (for example in a chain of accelerators), then single-turn extraction may be performed analogously to injection. Resonant extraction may also be employed.

!A storage ring can be a very useful component of a flying ship.!

This ship (UFO) has two circular accelerators (see the below image, fig. 122).

One accelerator is used permanently and the second can be used only at high loads. It works generally like a storage ring.

Fig. 122. This ship (UFO) has two circular accelerators.

Electron synchrotron

Circular electron accelerators fell somewhat out of favor for particle physics around the time that SLAC was constructed, because their synchrotron losses were considered economically prohibitive and because their beam intensity was lower than for the unpulsed linear machines. The Cornell Electron Synchrotron, built at low cost in the late 1960s, was the first in a series of high-energy circular electron accelerators built for fundamental particle physics, culminating in the LEP at CERN.

A large number of electron synchrotrons have been built in the past two decades, specialized to be synchrotron light sources, of ultraviolet light and X rays.

Synchrotron radiation sources

Some circular accelerators have been built to deliberately generate radiation (called synchrotron light) as X-rays also called synchrotron radiation, for example the Diamond Light Source which has been built at the Rutherford Appleton Laboratory in England or the Advanced Photon Source at Argonne National Laboratory in Illinois, USA. High-energy X-rays are useful for X-ray spectroscopy of proteins or X-ray absorption fine structure (XAFS) for example (fig. 123).

Synchrotron radiation is more powerfully emitted by lighter particles, so these accelerators are invariably electron accelerators. Synchrotron radiation allows for better imaging as researched and developed at SLAC's SPEAR.

A synchrotron light source is a source of electromagnetic radiation produced by a synchrotron, which is artificially produced for scientific and technical purposes by specialized particle accelerators, typically accelerating electrons. Once the high-energy electron beam has been generated, it is directed into auxiliary components such as bending magnets and insertion devices (undulators or wigglers) in storage rings and free electron lasers. These supply the strong magnetic fields perpendicular to the beam which are needed to convert the high-energy electron energy into light or some other form of EM radiation.

Fig. 123. *Synchrotron radiation emerging from a beam port. The blue colour comes from oxygen and nitrogen atoms in the air, ionised by the X-rays.*

The major applications of synchrotron light are in condensed matter physics, materials science, biology and medicine. A large fraction of experiments using synchrotron light involve probing the structure of matter from the sub-nanometer level of electronic structure to the micrometer and millimeter level important in medical imaging. An example of a practical industrial application is the manufacturing of microstructures by the LIGA process.

Synchrotron radiation may occur in accelerators either as a nuisance, causing undesired energy loss in particle physics contexts, or as a deliberately produced radiation source for numerous laboratory applications. Electrons are accelerated to high speeds in several stages to achieve a final energy that is typically in the gigaelectronvolt range. The electrons are forced to travel in a closed path by strong magnetic fields. This is similar to a radio antenna, but with the difference that the relativistic speed changes the observed frequency due to the Doppler effect by a factor γ. Relativistic Lorentz contraction bumps the frequency by another factor of γ, thus multiplying the gigahertz frequency of the resonant cavity that accelerates the electrons into the X-ray range. Another dramatic effect of relativity is that the radiation pattern is distorted from the isotropic dipole pattern expected from non-relativistic theory into an extremely forward-pointing cone of radiation. This makes synchrotron radiation sources the brightest known sources of X-rays. The planar acceleration geometry makes the radiation linearly polarized when observed in the orbital plane, and circularly polarized when observed at a small angle to that plane.

The advantages of using synchrotron radiation for spectroscopy and diffraction have been realized by an ever-growing scientific community, beginning in the 1960s and 1970s. In the beginning, accelerators were built for particle physics, and synchrotron radiation was used in "parasitic mode" when bending magnet radiation had to be extracted by drilling extra holes in the beam pipes. The first storage ring commissioned as a synchrotron light source was Tantalus, at the Synchrotron Radiation Center, first operational in 1968. As accelerator synchrotron radiation became more intense and its applications more promising, devices that enhanced the intensity of synchrotron radiation were built into existing rings. Third-generation synchrotron radiation sources were conceived and optimized from the outset to produce bright X-rays. Fourth-generation sources that will include different concepts for producing ultrabright, pulsed time-structured X-rays for extremely demanding and also probably yet-to-be-conceived experiments are under consideration.

Bending electromagnets in the accelerators were first used to generate the radiation; but to generate stronger radiation, other specialized devices, called insertion devices, are sometimes employed. Current third-generation synchrotron radiation sources are typically heavily based upon these insertion devices, when straight sections in the storage ring are used for inserting periodic magnetic structures (composed of many magnets that have a special repeating row of N and S poles) that force the electrons into a sinusoidal path or helical path. Thus, instead of a single bend, many tens or hundreds of "wiggles" at precisely calculated positions add up or multiply the total intensity that is seen at the end of the straight section. These devices are called wigglers or undulators. The main difference between an undulator and a wiggler is the intensity of their magnetic field and the amplitude of the deviation from the straight line path of the electrons.

There are openings in the storage ring to let the radiation exit and follow a beam line into the experimenters' vacuum chamber. A great number of such beamlines can emerge from modern third-generation synchrotron radiation sources.

Beamlines

At a synchrotron facility, electrons are usually accelerated by a synchrotron, and then injected into a storage ring, in which they circulate, producing synchrotron radiation, but without gaining further energy.

The radiation is projected at a tangent to the electron storage ring and captured by beamlines. These beamlines may originate at bending magnets, which mark the corners of the storage ring; or insertion devices, which are located in the straight sections of the storage ring.

The spectrum and energy of X-rays differ between the two types. The beamline includes X-ray optical devices which control the bandwidth, photon flux, beam dimensions, focus, and collimation of the rays.

The optical devices include slits, attenuators, crystal monochromators, and mirrors. The mirrors may be bent into curves or toroidal shapes to focus the beam. A high photon flux in a small area is the most common requirement of a beamline.

The design of the beamline will vary with the application. At the end of the beamline is the experimental end station, where samples are placed in the line of the radiation, and detectors are positioned to measure the resulting diffraction, scattering or secondary radiation.

Experimental techniques and usage

Synchrotron light is an ideal tool for many types of research and also has industrial applications. Some of the experimental techniques in synchrotron beamlines are:

Structural analysis of crystalline and amorphous materials

Powder diffraction analysis

Crystallography of proteins and other macromolecules

Magnetic scattering

Small angle X-ray scattering

X-ray absorption spectroscopy

Inelastic X-ray scattering

Soft X-ray emission spectroscopy

Tomography

X-ray imaging in phase contrast mode

X-ray standing wave experiments

Photolithography for MEMS structures as part of the LIGA process.

High pressure studies

residual stress analysis

X-Ray Multiple Diffraction

Photoemission spectroscopy and Angle resolved photoemission spectroscopy

Some of the advantages of synchrotron light that allow for these practical uses are:

High energy X-rays, short wavelength photons which can penetrate matter and interact with atoms.

High concentration, tunability and polarization thus ensuring focusing accuracy for even the smallest of targets.

Compact synchrotron light sources

Because of the usefulness of tuneable collimated coherent X-ray radiation, efforts have been made to make smaller more economical sources of the light produced by synchrotrons.

The aim is to make such sources available within a research laboratory for cost and convenience reasons; at present, researchers have to travel to a facility to perform experiments.

One method of making a compact light source is to utilise the energy shift from Compton scattering near-visible laser photons from electrons stored at relatively low energies of tens of megaelectronvolts (see for example the Compact Light Source (CLS)).

However, a relatively low cross-section of collision can be obtained in this manner, and the repetition rate of the lasers is limited to a few hertz rather than the megahertz repetition rates naturally arising in normal storage ring emission.

Another method is to use plasma acceleration to reduce the distance required to accelerated electrons from rest to the energies required for UV or X-ray emission within magnetic devices.

Synchrotron light

Synchrotron light (also known as synchrotron radiation) is electromagnetic radiation that is emitted when charged particles moving at close to the speed of light are forced to change direction by a magnetic field. Synchrotron light can be produced naturally by astronomical objects, such as the Crab Nebula – a supernova remnant in the Taurus constellation. Since the late 1940s synchrotron light has been artificially generated using synchrotrons – particle accelerators that gave the phenomenon its name.

Synchrotron radiation spans a wide frequency range, from infrared up to the highest-energy X-rays. It is characterised by high brightness – many orders of magnitude brighter than conventional sources – and the light is highly polarised, tunable, collimated (consisting of almost parallel rays) and concentrated over a small area.

When synchrotrons were first developed, their primary purpose was to accelerate particles for the study of the nucleus, not to generate light.

Today on the other hand, while a few are still used as colliders for high-energy physics experiments such as the Large Hadron Collider at CERN, there are more than 50 synchrotron light sources around the world dedicated to generating synchrotron light and exploiting its special qualities. These machines support a huge range of applications, from condensed matter physics to structural biology, environmental science and cultural heritage.

Earlier accelerators, called cyclotrons, had fixed magnetic fields. Because the bending of a charged particle is inversely proportional to its momentum, cyclotrons were limited to fairly low energies otherwise they became unaffordably large. By collecting the particles into bunches and synchronising a rise in magnetic-field strength with the increasing energy of the charged particles, the particles could be accelerated to higher energies while being constrained to a fixed circular path. Thus the synchrotron was born, with the first observation of artificial synchrotron light occurring at General Electric in the US in 1947.

British theoretical physicist James Maxwell's classical theory of electromagnetism (first fully documented in 1873) explains that charged particles moving through a magnetic field generate electromagnetic radiation. By requiring Maxwell's equations to be true in all inertial (non-accelerating) frames of reference, Einstein's theory of special relativity fully explained the complete characteristics of synchrotron light generated by electrons travelling along a circular path at relativistic speeds. The radiation produced in synchrotrons is focused in a narrow cone, perpendicular to its acceleration direction and parallel to the directional motion of the electrons.

Synchrotron light is the brightest artificial source of X-rays, allowing the detailed study of molecular structures, which has led to the award of Nobel prizes in a number of fields (see timeline).

In 1956, two American scientists, Diran Tomboulian and Paul Hartman, were granted use of the 320 MeV synchrotron at Cornell University. In addition to confirming the spectral and angular distribution of synchrotron light, they carried out the first X-ray spectroscopy study using synchrotron light. Five years later, the National Bureau of Standards in the US modified its 180 MeV machine to allow synchrotron light to be harvested for experiments. These became known as the first-generation synchrotrons – machines built for smashing nuclei apart using electrons, which were later used for synchrotron-light experiments.

Over the coming decades, as demand grew for the use of first generation synchrotron machines, pioneering advances led to a number of developments, such as the storage ring, which allowed particles to circulate for long periods of time providing more stable beam conditions, benefiting both particle physicists and synchrotron users. One of the most significant developments took place in the late 1970s when plans were approved to build the world's first dedicated synchrotron light source producing X-rays (Synchrotron Radiation Source – SRS) at Daresbury in the UK, which started user experiments in 1981. The US, Japan and others also built second-generation machines, while other first-generation machines received upgrades to allow for more experiments.

For scientists carrying out spectroscopy experiments, the brightness of the beam reaching the sample determined the resolving power of the results. For crystallographers, especially those looking at small crystals with large unit cells, high brightness was important to resolve closely spaced diffraction spots. As second-generation machines were optimised to produce brighter beams, a fundamental limit was approaching. To meet the increasing demands of a growing synchrotron user community, a new approach was required: insertion devices.

Insertion devices are arrays of magnets placed into the straight sections of the storage ring, which could be retro-fitted to second-generation machines and were quickly incorporated into existing synchrotrons. Insertion devices help to create a beam that is very bright and with intensity peaks with a wavelength that can be varied by adjusting the field strength (often the gap between two magnet arrays).

The increased brightness made data collection faster, and tunable wavelengths benefited crystallographers and spectroscopists alike. By the early 1990s machines were being designed with insertion devices in place from the start, and the first such third-generation source, the European Synchrotron Radiation Facility (ESRF) in Grenoble, France, started operating in 1994. There are now more than 50 dedicated light sources in the world, combining both second- and third-generation machines, which cover a wide spectral range from infrared to hard X-rays.

The experimental configurations of different synchrotron facilities are quite similar. The storage rings where the light is generated have many ports, which each open onto a beamline, where scientists set up their experiments and collect data. The beamlines, however, can vary a lot in the details depending on the experimental methods they are used for. The Diamond Light Source in Oxfordshire became Britain's newest synchrotron light facility in 2007 and in August 2008 it was Britain's only synchrotron source following the closure of the SRS. It represents the largest single UK science investment for 40 years and will have the capacity to host 40 beamlines. The Diamond Light Source supports a huge range of scientific disciplines, including condensed matter physics, chemistry, nanophysics, structural biology, engineering, environmental science and cultural heritage.

Synchrotron X-ray beams allow detailed analysis and modelling of strain, cracks and corrosion as well as in situ study of materials during production processing. This research is vital to the development of high-performance materials and their use in innovative products and structures. The Diamond Light Source has been used to study the processes behind pitting corrosion, which attacks the so-called corrosion-resistant metals used in containers for nuclear waste, and to understand how applied stresses can cause cracks to propagate through materials.

Synchrotron-based techniques have made a major impact in the field of environmental science in the last 10 years. High brightness allows high-resolution study of ultra-dilute substances, the identification of species and the ability to track pollutants as they move through the environment. Synchrotrons have been used to develop more efficient techniques for hydrogen storage and to study the way in which depleted uranium disperses into the local environment. Tiny heavy-metal samples excreted from earthworms have been compared with contaminated soil samples, revealing how earthworms survive in these environments and introducing the idea that earthworms could help to decontaminate land.

Determining the properties and morphology of buried layers and interfaces is an important area in solid-state science with synchrotrons being the meeting ground of state-of-the-art theory and high-precision experimental results.

Many of the technological products of materials science are based on thin-film devices, which consist of a series of such layers. Structural studies of in situ processing of semiconducting polymer films are also likely to be an important area of growth in the coming decade.

Diffraction of high-intensity X-ray beams is an ideal technique to study spin, charge and orbital ordering in single-crystal samples to understand high-temperature superconductivity. The SRS was used to help study giant magneto-resistance (GMR), which is now used in billions of electronic devices worldwide.

The demand for synchrotron light has meant that third-generation machines are being built around the world, and existing machines continue to be developed to provide brighter X-rays, increased user hours and more flexible experimental stations.

The modular nature of modern synchrotrons means that new technologies can be incorporated into existing machines as they arrive. By using powerful linear accelerator technology, fourth-generation sources – known as free-electron lasers (FELs) – can generate shorter, femtosecond pulses but with the same intensity in each peak as synchrotron sources emit in one second, producing X-rays that are millions of times brighter in each pulse than the most powerful synchrotrons. FELs won't replace third-generation machines, but will provide facilities that enable studies at higher peak brightness.

Collimated light

Collimated light is light whose rays are nearly parallel, and therefore will spread slowly as it propagates. The word is related to "collinear" and implies light that does not disperse with distance (ideally), or that will disperse minimally (in reality). A perfectly collimated beam with no divergence cannot be created due to

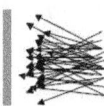

diffraction, but light can be approximately collimated by a number of processes, for instance by means of a collimator. Collimated light is sometimes said to be focused at infinity. Thus as the distance from a point source increases, the

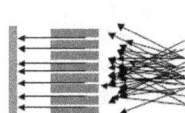

spherical wavefronts become flatter and closer to plane waves, which are perfectly collimated (fig. 124).

Fig. 124. *In the lower picture,* *the light has been collimated.*

Synchrotrons produce a unique type of radiation—continuous across the spectrum and tunable to the desired wavelength—emitted by electrons accelerated in a magnetic field.

For two decades, the Office of Science has been the major supporter of U.S. synchrotron light sources.

It currently operates four, each with unique capabilities, used by a total of more than 6,000 researchers annually from academia, government, and industry.

The four are the Advanced Light Source at Lawrence Berkeley National Laboratory, Advanced Photon Source at Argonne National Laboratory, National Synchrotron Light Source at Brookhaven

National Laboratory, and Stanford Synchrotron Radiation Laboratory at Stanford Linear Accelerator Center.

Scientists at these sites helped pioneer many synchrotron innovations that are widely used today, including a lattice of magnets that increased brightness (photon density) by two orders of magnitude; "insertion" devices (linear arrays of magnets called wigglers and undulators) that oscillate the path of the electron beam to generate X-ray and ultraviolet light that is high in flux (number of photons) and collimation (parallel alignment of photons); and powerful experimental techniques such as X-ray scattering and X-ray microscopy.

The "collimated light" obtained in a synchrotron, can be used to move flying ships.

Energy and momentum (of the evacuated light) must be very large.

We can increase the number of pulses per second and number of particles per pulse, but we can not increase the speed of a particle. We already work with the greatest velocity experimentally determined, the speed of light.

However we can increase the photons energy by increasing the frequency the beam light.

The modular nature of modern synchrotrons means that new technologies can be incorporated into existing machines as they arrive.

By using powerful linear accelerator technology, fourth-generation sources – known as free-electron lasers (FELs) – can generate shorter, femtosecond pulses but with the same intensity in each peak as synchrotron sources emit in one second, producing X-rays that are millions of times brighter in each pulse than the most powerful synchrotrons.

A synchrotron light source is a source of electromagnetic radiation produced by a synchrotron, which is artificially produced for scientific and technical purposes by specialized particle accelerators, typically accelerating electrons.

Once the high-energy electron beam has been generated, it is directed into auxiliary components such as bending magnets and insertion devices (undulators or wigglers) in storage rings and free electron lasers.

These supply the strong magnetic fields perpendicular to the beam which are needed to convert the high-energy electron energy into light or some other form of EM radiation.

Any ship engine needs energy (power) and fuel.

A ship with a beam engine needs only energy (power). The ship's fuel is represented by the electrons. Energy (power) (which produces and the electrons) gives also and the fuel role.

Such a ship requires only a nuclear reactor which donates its energy (power). The power can be completed with some solar panels or with another energy source. Without fuel! The ship autonomy increase very much.

An ion thruster can't take off or land the ship in an gravitational field. Not even on moon!

We need very, very high energies! The Super LASER can do that. But today and especially tomorrow, we can produce higher energies with accelerators, with particle accelerators!

We can use and linear accelerators. Or a combination between a circular accelerator and a linear accelerator.

Or we will accelerate a particle up to 10-100 PeV, or it increases the flow of particles per pulse and the pulses per second.

LASERS

A laser is a device that emits light (electromagnetic radiation) through a process of optical amplification based on the stimulated emission of photons.

The term "laser" originated as an acronym for Light Amplification by Stimulated Emission of Radiation.

The emitted laser light is notable for its high degree of spatial and temporal coherence, unattainable using other technologies.

Spatial coherence typically is expressed through the output being a narrow beam which is diffraction-limited, often a so-called "pencil beam."

Laser beams can be focused to very tiny spots, achieving a very high irradiance. Or they can be launched into a beam of very low divergence in order to concentrate their power at a large distance.

Temporal (or longitudinal) coherence implies a polarized wave at a single frequency whose phase is correlated over a relatively large distance (the coherence length) along the beam.

A beam produced by a thermal or other incoherent light source has an instantaneous amplitude and phase which vary randomly with respect to time and position, and thus a very short coherence length.

Most so-called "single wavelength" lasers actually produce radiation in several modes having slightly different frequencies (wavelengths), often not in a single polarization.

And although temporal coherence implies monochromaticity, there are even lasers that emit a broad spectrum of light, or emit different wavelengths of light simultaneously.

There are some lasers which are not single spatial mode and consequently their light beams diverge more than required by the diffraction limit. However all such devices are classified as "lasers" based on their method of producing that light: stimulated emission. Lasers are employed in applications where light of the required spatial or temporal coherence could not be produced using simpler technologies.

The word laser started as an acronym for "light amplification by stimulated emission of radiation"; in modern usage "light" broadly denotes electromagnetic radiation of any frequency, not only visible light, hence infrared laser, ultraviolet laser, X-ray laser, and so on.

Because the microwave predecessor of the laser, the maser, was developed first, devices of this sort operating at microwave and radio frequencies are referred to as "masers" rather than "microwave lasers" or "radio lasers" (fig. 125).

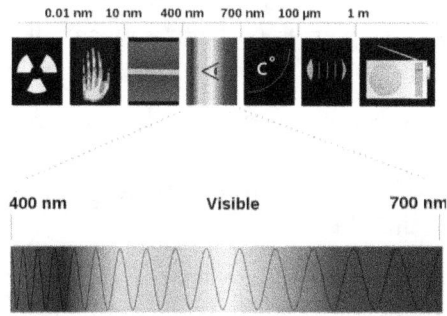

Fig. 125. *From left to right: gamma rays, X-rays, ultraviolet rays, visible spectrum, infrared, microwaves, radio waves. Bottom: enlargement of visible spectrum from violet (400nm) to red (700nm).*

In the early technical literature, especially at Bell Telephone Laboratories, the laser was called an optical maser; this term is now obsolete.

A laser which produces light by itself is technically an optical oscillator rather than an optical amplifier as suggested by the acronym.

It has been humorously noted that the acronym LOSER, for "light oscillation by stimulated emission of radiation," would have been more correct.

With the widespread use of the original acronym as a common noun, actual optical amplifiers have come to be referred to as "laser amplifiers", notwithstanding the apparent redundancy in that designation.

The back-formed verb to lase is frequently used in the field, meaning "to produce laser light," especially in reference to the gain medium of a laser; when a laser is operating it is said to be "lasing." Further use of the words laser and maser in an extended sense, not referring to laser technology or devices, can be seen in usages such as astrophysical maser and atom laser.

A laser consists of a gain medium inside a highly reflective optical cavity, as well as a means to supply energy to the gain medium. The gain medium is a material with properties that allow it to amplify light by stimulated emission. In its simplest form, a cavity consists of two mirrors arranged such that light bounces back and forth, each time passing through the gain medium. Typically one of the two mirrors, the output coupler, is partially transparent. The output laser beam is emitted through this mirror.

Light of a specific wavelength that passes through the gain medium is amplified (increases in power); the surrounding mirrors ensure that most of the light makes many passes through the gain medium, being amplified repeatedly.

Part of the light that is between the mirrors (that is, within the cavity) passes through the partially transparent mirror and escapes as a beam of light.

The process of supplying the energy required for the amplification is called pumping. The energy is typically supplied as an electrical current or as light at a different wavelength. Such light may be provided by a flash lamp or perhaps another laser.

Most practical lasers contain additional elements that affect properties such as the wavelength of the emitted light and the shape of the beam.

The gain medium of a laser is a material of controlled purity, size, concentration, and shape, which amplifies the beam by the process of stimulated emission. It can be of any state: gas, liquid, solid or plasma.

The gain medium absorbs pump energy, which raises some electrons into higher-energy ("excited") quantum states. Particles can interact with light by either absorbing or emitting photons.

Emission can be spontaneous or stimulated. In the latter case, the photon is emitted in the same direction as the light that is passing by.

When the number of particles in one excited state exceeds the number of particles in some lower-energy state, population inversion is achieved and the amount of stimulated emission due to light that passes through is larger than the amount of absorption.

Hence, the light is amplified. By itself, this makes an optical amplifier. When an optical amplifier is placed inside a resonant optical cavity, one obtains a laser.

The light generated by stimulated emission is very similar to the input signal in terms of wavelength, phase, and polarization. This gives laser light its characteristic coherence, and allows it to maintain the uniform polarization and often monochromaticity established by the optical cavity design.

The optical resonator is sometimes referred to as an "optical cavity", but this is a misnomer: lasers use open resonators as opposed to the literal cavity that would be employed at microwave frequencies in a maser.

The resonator typically consists of two mirrors between which a coherent beam of light travels in both directions, reflecting back on itself so that an average photon will pass through the gain medium repeatedly before it is emitted from the output aperture or lost to diffraction or absorption.

If the gain (amplification) in the medium is larger than the resonator losses, then the power of the recirculating light can rise exponentially. But each stimulated emission event returns an atom from its excited state to the ground state, reducing the gain of the medium.

With increasing beam power the net gain (gain times loss) reduces to unity and the gain medium is said to be saturated. In a continuous wave (CW) laser, the balance of pump power against gain saturation and cavity losses produces an equilibrium value of the laser power inside the cavity; this equilibrium determines the operating point of the laser. If the applied pump power is too small, the gain will never be sufficient to overcome the resonator losses, and laser light will not be produced.

The minimum pump power needed to begin laser action is called the lasing threshold. The gain medium will amplify any photons passing through it, regardless of direction; but only the photons in a spatial mode supported by the resonator will pass more than once through the medium and receive substantial amplification.

The beam in the cavity and the output beam of the laser, when travelling in free space (or a homogenous medium) rather than waveguides (as in an optical fiber laser), can be approximated as a Gaussian beam in most lasers; such beams exhibit the minimum divergence for a given diameter.

However some high power lasers may be multimode, with the transverse modes often approximated using Hermite-Gaussian or Laguerre-Gaussian functions. It has been shown that unstable laser resonators (not used in most lasers) produce fractal shaped beams. Near the beam "waist" (or focal region) it is highly collimated: the wavefronts are planar, normal to the direction of propagation, with no beam divergence at that point.

However due to diffraction, that can only remain true well within the Rayleigh range. The beam of a single transverse mode (gaussian beam) laser eventually diverges at an angle which varies inversely with the beam diameter, as required by diffraction theory. Thus, the "pencil beam" directly generated by a common helium-neon laser would spread out to a size of perhaps 500 kilometers when shone on the Moon (from the distance of the earth). On the other hand the light from a semiconductor laser typically exits the tiny crystal with a large divergence: up to 50°.

However even such a divergent beam can be transformed into a similarly collimated beam by means of a lens system, as is always included, for instance, in a laser pointer whose light originates from a laser diode. That is possible due to the light being of a single spatial mode.

This unique property of laser light, spatial coherence, cannot be replicated using standard light sources (except by discarding most of the light) as can be appreciated by comparing the beam from a flashlight (torch) or spotlight to that of almost any laser.

The mechanism of producing radiation in a laser relies on stimulated emission, where energy is extracted from a transition in an atom or molecule.

This is a quantum phenomenon discovered by Einstein who derived the relationship between the A coefficient describing spontaneous emission and the B coefficient which applies to absorption and stimulated emission.

However in the case of the free electron laser, atomic energy levels are not involved; it appears that the operation of this rather exotic device can be explained without reference to quantum mechanics.

A laser can be classified as operating in either continuous or pulsed mode, depending on whether the power output is essentially continuous over time or whether its output takes the form of pulses of light on one or another time scale.

Of course even a laser whose output is normally continuous can be intentionally turned on and off at some rate in order to create pulses of light.

When the modulation rate is on time scales much slower than the cavity lifetime and the time period over which energy can be stored in the lasing medium or pumping mechanism, then it is still classified as a "modulated" or "pulsed" continuous wave laser. Most laser diodes used in communication systems fall in that category.

Some applications of lasers depend on a beam whose output power is constant over time. Such a laser is known as continuous wave (CW). Many types of lasers can be made to operate in continuous wave mode to satisfy such an application.

Many of these lasers actually lase in several longitudinal modes at the same time, and beats between the slightly different optical frequencies of those oscillations will in fact produce amplitude variations on time scales shorter than the round-trip time (the reciprocal of the frequency spacing between modes), typically a few nanoseconds or less.

In most cases these lasers are still termed "continuous wave" as their output power is steady when averaged over any longer time periods, with the very high frequency power variations having little or no impact in the intended application. (However the term is not applied to mode locked lasers, where the intention is to create very short pulses at the rate of the round-trip time).

For continuous wave operation it is required for the population inversion of the gain medium to be continually replenished by a steady pump source.

In some lasing media this is impossible. In some other lasers it would require pumping the laser at a very high continuous power level which would be impractical or destroy the laser by producing excessive heat. Such lasers cannot be run in CW mode.

Pulsed operation of lasers refers to any laser not classified as continuous wave, so that the optical power appears in pulses of some duration at some repetition rate. This encompasses a wide range of technologies addressing a number of different motivations. Some lasers are pulsed simply because they cannot be run in continuous mode.

In other cases the application requires the production of pulses having as large an energy as possible. Since the pulse energy is equal to the average power divided by the repetition rate, this goal can sometimes be satisfied by lowering the rate of pulses so that more energy can be built up in between pulses. In laser ablation for example, a small volume of material at the surface of a work piece can be evaporated if it is heated in a very short time, whereas supplying the energy gradually would allow for the heat to be absorbed into the bulk of the piece, never attaining a sufficiently high temperature at a particular point.

Other applications rely on the peak pulse power (rather than the energy in the pulse), especially in order to obtain nonlinear optical effects. For a given pulse energy, this requires creating pulses of the shortest possible duration utilizing techniques such as Q-switching.

The optical bandwidth of a pulse cannot be narrower than the reciprocal of the pulse width. In the case of extremely short pulses, that implies lasing over a considerable bandwidth, quite contrary to the very narrow bandwidths typical of CW lasers. The lasing medium in some dye lasers and vibronic solid-state lasers produces optical gain over a wide bandwidth, making a laser possible which can thus generate pulses of light as short as a few femtoseconds (10^{-15} s).

!Laser for propulsion has the great advantage to operate with very many pulses per second (high frequency).!

In a Q-switched laser, the population inversion is allowed to build up by introducing loss inside the resonator which exceeds the gain of the medium; this can also be described as a reduction of the quality factor or 'Q' of the cavity. Then, after the pump energy stored in the laser medium has approached the maximum possible level, the introduced loss mechanism (often an electro- or acousto-optical element) is rapidly removed (or that occurs by itself in a passive device), allowing lasing to begin which rapidly obtains the stored energy in the gain medium. This results in a short pulse incorporating that energy, and thus a high peak power.

A mode-locked laser is capable of emitting extremely short pulses on the order of tens of picoseconds down to less than 10 femtoseconds. These pulses will repeat at the round trip time, that is, the time that it takes light to complete one round trip between the mirrors comprising the resonator. Due to the Fourier limit (also known as energy-time uncertainty), a pulse of such short temporal length has a spectrum spread over a considerable bandwidth. Thus such a gain medium must have a gain bandwidth sufficiently broad to amplify those frequencies. An example of a suitable material is titanium-doped, artificially grown sapphire (Ti:sapphire) which has a very wide gain bandwidth and can thus produce pulses of only a few femtoseconds duration.

Such mode-locked lasers are a most versatile tool for researching processes occurring on extremely short time scales (known as femtosecond physics, femtosecond chemistry and ultrafast science), for maximizing the effect of nonlinearity in optical materials (e.g. in second-harmonic generation, parametric down-conversion, optical parametric oscillators and the like) due to the large peak power, and in ablation applications. Again, because of the extremely short pulse duration, such a laser will produce pulses which achieve an extremely high peak power.

Another method of achieving pulsed laser operation is to pump the laser material with a source that is itself pulsed, either through electronic charging in the case of flash lamps, or another laser which is already pulsed. Pulsed pumping was historically used with dye lasers where the inverted population lifetime of a dye molecule was so short that a high energy, fast pump was needed. The way to overcome this problem was to charge up large capacitors which are then switched to discharge through flashlamps, producing an intense flash.

Pulsed pumping is also required for three-level lasers in which the lower energy level rapidly becomes highly populated preventing further lasing until those atoms relax to the ground state. These lasers, such as the excimer laser and the copper vapor laser, can never be operated in CW mode.

In 1917, Albert Einstein established the theoretic foundations for the laser and the maser in the paper Zur Quantentheorie der Strahlung (On the Quantum Theory of Radiation); via a re-derivation of Max Planck's law of radiation, conceptually based upon probability coefficients (Einstein coefficients) for the absorption, spontaneous emission, and stimulated emission of electromagnetic radiation; in 1928, Rudolf W. Ladenburg confirmed the existences of the phenomena of stimulated emission and negative absorption; in 1939, Valentin A. Fabrikant predicted the use of stimulated emission to amplify "short" waves; in 1947, Willis E. Lamb and R. C. Retherford found apparent stimulated emission in hydrogen spectra and effected the first demonstration of stimulated emission; in 1950, Alfred Kastler (Nobel Prize for Physics 1966) proposed the method of optical pumping, experimentally confirmed, two years later, by Brossel, Kastler, and Winter.

In 1953, Charles Hard Townes and graduate students James P. Gordon and Herbert J. Zeiger produced the first microwave amplifier, a device operating on similar principles to the laser, but amplifying microwave radiation rather than infrared or visible radiation. Townes's maser was incapable of continuous output. Meanwhile, in the Soviet Union, Nikolay Basov and Aleksandr Prokhorov were independently working on the quantum oscillator and solved the problem of continuous-output systems by using more than two energy levels. These gain media could release stimulated emissions between an excited state and a lower excited state, not the ground state, facilitating the maintenance of a population inversion. In 1955, Prokhorov and Basov suggested optical pumping of a multi-level system as a method for obtaining the population inversion, later a main method of laser pumping.

Townes reports that several eminent physicists — among them Niels Bohr, John von Neumann, Isidor Rabi, Polykarp Kusch, and Llewellyn Thomas — argued the maser violated Heisenberg's uncertainty principle and hence could not work. In 1964 Charles H. Townes, Nikolay Basov, and Aleksandr Prokhorov shared the Nobel Prize in Physics, "for fundamental work in the field of quantum electronics, which has led to the construction of oscillators and amplifiers based on the maser–laser principle".

In 1957, Charles Hard Townes and Arthur Leonard Schawlow, then at Bell Labs, began a serious study of the infrared laser. As ideas developed, they abandoned infrared radiation to instead concentrate upon visible light. The concept originally was called an "optical maser". In 1958, Bell Labs filed a patent application for their proposed optical maser; and Schawlow and Townes submitted a manuscript of their theoretical calculations to the Physical Review, published that year in Volume 112, Issue No. 6.

Simultaneously, at Columbia University, graduate student Gordon Gould was working on a doctoral thesis about the energy levels of excited thallium. When Gould and Townes met, they spoke of radiation emission, as a general subject; afterwards, in November 1957, Gould noted his ideas for a "laser", including using an open resonator (later an essential laser-device component).

Moreover, in 1958, Prokhorov independently proposed using an open resonator, the first published appearance (the USSR) of this idea. Elsewhere, in the US, Schawlow and Townes had agreed to an open-resonator laser design — apparently unaware of Prokhorov's publications and Gould's unpublished laser work.

At a conference in 1959, Gordon Gould published the term LASER in the paper The LASER, Light Amplification by Stimulated Emission of Radiation. Gould's linguistic intention was using the "-aser" word particle as a suffix — to accurately denote the spectrum of the light emitted by the LASER device; thus x-rays: xaser, ultraviolet: uvaser, et cetera; none established itself as a discrete term, although "raser" was briefly popular for denoting radio-frequency-emitting devices.

Gould's notes included possible applications for a laser, such as spectrometry, interferometry, radar, and nuclear fusion. He continued developing the idea, and filed a patent application in April 1959. The U.S. Patent Office denied his application, and awarded a patent to Bell Labs, in 1960.

That provoked a twenty-eight-year lawsuit, featuring scientific prestige and money as the stakes. Gould won his first minor patent in 1977, yet it was not until 1987 that he won the first significant patent lawsuit victory, when a Federal judge ordered the US Patent Office to issue patents to Gould for the optically pumped and the gas discharge laser devices.

Since the early period of laser history, laser research has produced a variety of improved and specialized laser types, optimized for different performance goals, including:

new wavelength bands

maximum average output power

maximum peak pulse energy

maximum peak pulse power

minimum output pulse duration

maximum power efficiency

minimum cost

And this research continues to this day.

List of laser types

Gas lasers, Chemical lasers, Excimer lasers, Solid-state lasers, Fiber lasers, Photonic crystal lasers, Semiconductor lasers, Dye lasers, Free electron lasers, Biolaser, and others.

Positronium annihilation laser:

In September 2007, the BBC News reported that there was speculation about the possibility of using positronium annihilation to drive a very powerful gamma ray laser. Dr. David Cassidy of the University of California, Riverside proposed that a single such laser could be used to ignite a nuclear fusion reaction, replacing the banks of hundreds of lasers currently employed in inertial confinement fusion experiments.

Space-based X-ray lasers pumped by a nuclear explosion have also been proposed as antimissile weapons. Such devices would be one-shot weapons.

REFERENCES

[1]-"Air Force World: B-2 Crash Cause Identified", AIR FORCE Magazine, July 2008, Vol. 91, No.7, pp. 16-17;

[2]-Archytas of Tarentum, Technology Museum of Thessaloniki, Macedonia, Greece, retrieved 2009-06-27;

[3]-Barde, Robert E. "Midway: Tarnished Victory", Military Affairs, v. 47, no. 4 (December 1983);

[4]-Bergerund, Eric M. (2000). Fire in the Sky: The Air War in the South Pacific. Boulder, Colorado: Westview Press. p. 752. ISBN 978-0-8133-2985-7;

[5]-Bess, Michael (2006). Choices Under Fire: Moral Dimensions of World War II. New York: Alfred A. Knopf. ISBN 0-307-26365-7;

[6]-Bicheno, Hugh. Midway. London: Orion Publishing Group, 2001 (reprints Cassell 2001 edition);

[7]-Blair Jr., Clay (1975). Silent Victory: The U.S. Submarine War Against Japan. Philadelphia: J.B. Lippincott. p. 1072;

[8]-Buell, Thomas B. (1987). The Quiet Warrior: a Biography of Admiral Raymond A. Spruance. Annapolis, Md.: Naval Institute Press. p. 518. ISBN 0-87021-562-0;

[9]-Chesneau, Roger (ed.) (1980). Conway's All the World's Fighting Ships 1922–1946. London: Conway Maritime Press. ISBN 0-85177-146-7;

[10]-Cressman, Robert J.; et.al. (1990). "A Glorious page in our history," Adm. Chester Nimitz, 1942: the Battle of Midway, 4–6 June 1942. Missoula, Mont.: Pictorial Histories Pub. Co.. ISBN 0-929521-40-4.

[11]-Crouch, Tom (2004), Wings: A History of Aviation from Kites to the Space Age, New York, New York: W.W. Norton & Co;

[12]-Darling, David J. (2003), The Complete Book of Spaceflight: From Apollo 1 to zero Gravity, New York, New York: John Wiley and Sons, ISBN 0-471-05649-9, retrieved 2010-09-13;

[13]-Deng, Yinke; Wang, Pingxing (2005), Ancient Chinese Inventions, China Intercontinental Press, pp. 113, ISBN 7-5085-0837-8, retrieved 2010-09-13;

[14]-Dodson, MG (2005), "An Historical and Applied Aerodynamic Study of the Wright Brothers' Wind Tunnel Test Program and Application to Successful Manned Flight", US Naval Academy Technical Report USNA-334, retrieved 2009-03-11;

[15]-Dreams of Leonardo, program by Public Broadcasting Service (PBS), October 2005, describes the building and successful flight of a glider based on Leonardo's design;

[16]-Evans, David; Peattie, Mark R. (1997). Kaigun: Strategy, Tactics, and Technology in the Imperial Japanese Navy, 1887–1941. Annapolis, Maryland: Naval Institute Press. ISBN 0-87021-192-7;

[17]-Ewing, Steve (2004). Thach Weave: The Life of Jimmie Thach. Annapolis, Maryland: Naval Institute Press. ISBN 1-59114-248-2;

[18]-Ferdinando Pedriali. "Aerei italiani in Libia (1911–1912)"(Italian planes in Libya (1911–1912)). Storia Militare (Military History), N° 170/novembre 2007,p.31–40;

[19]-Frank H. Wenham, inventor of the wind tunnel, 1871, was a fan, driven by a steam engine, propelled air down a 12 ft (3.7 m). (3.7 m) tube to the model. NASA;

[20]-Gibbs-Smith, Charles Harvard (1960), The Aeroplane: An Historical Survey of Its Origins and Development, London: Her Majesty's Stationery Office, pp. 207–208;

[21]-Hakim, Joy (1995). A History of Us: War, Peace and all that Jazz. New York: Oxford University Press. ISBN 0-19-509514-6;

[22]-Holmes, W. (1979). Double-Edged Secrets: U.S. Naval Intelligence Operations in the Pacific During World War II (Bluejacket Books). Naval Institute Press. ISBN 1-55750-324-9;

[23]-Howe, Henry (1858), Memoirs of the most eminent American mechanics, New York, New York: Derby & Jackson, retrieved 2010-09-13;

[24]-http://en.wikipedia.org/wiki/STOVL;

[25]-Kelly, Fred C. The Wright Brothers: A Biography Chp. IV, p.101–102 (Dover Publications, NY 1943);-Abzug, Malcolm J. and E. Eugene Larrabee."Airplane Stability and Control, Second Edition: A History of the Technologies That Made Aviation Possible." cambridge.org. Retrieved: September 21, 2010;

[26]-Kulawik, Piotr. "Jan Wnęk l'héros de sous la voûte de ciel [Jan Wnęk, the hero in the vault of heaven]" (in French). Retrieved 2010-09-13;

[27]-Layton, Rear Admiral Edwin T. (1985). And I Was There: Pearl Harbor and Midway, Konecky and Konecky;

[28]-Lord, Walter (1967). Incredible Victory. Burford. ISBN 1-58080-059-9. Focuses primarily on the human experience of the battle;

[29]-Lundstrom, John B. (2005 (new edition)). The First Team: Pacific Naval Air Combat from Pearl Harbor to Midway. Annapolis, Maryland: Naval Institute Press. ISBN 1-59114-471-X;

[30]-Lynn Townsend White, Jr. (Spring, 1961). "Eilmer of Malmesbury, an Eleventh Century Aviator: A Case Study of Technological Innovation, Its Context and Tradition", Technology and Culture 2 (2), p. 97-111;

[31]-NASA- http://www.nasa.gov/;

[32]-Nevin, David. "Two Daring Flyers Beat the Atlantic before Lindbergh." Journal of Contemporary History' 28: (1) 1993, 105;

[33]-Parillo, Mark. Japanese Merchant Marine in World War II. Annapolis, Maryland: United States Naval Institute Press, 1993;

[34]-Parshall, Jonathan; Tully, Anthony (2005). Shattered Sword: The Untold Story of the Battle of Midway. Dulles, Virginia: Potomac Books. ISBN 1-57488-923-0. Uses recent Japanese sources;

[35]-Peattie, Mark R.. Sunburst: The Rise of Japanese Naval Air Power, 1909–1941. US Naval Institute Press. p. 392. ISBN 1-59114-664-X;

[36]-Petrescu, F. New Aircraft. COMEC 2009, Braşov, ROMANIA, 2009;
[37]-Robinson, Tim. "Not so hidden dragon – China's J-20 assessed." Aerospace Insight, 14 January 2011;

[38]-Smith, Michael (2000). The Emperor's Codes: Bletchley Park and the breaking of Japan's secret ciphers, Bantam Press, ISBN 0-593-04642-0. Chapter 11: "Midway: The battle that turned the tide";

[39]-Stephan, John J. (1984). Hawaii Under the Rising Sun: Japan's Plans for Conquest after Pearl Harbor. Honolulu: University of Hawaii Press. ISBN 0-8248-2550-0;

[40]-Stille, Mark (2007). USN Carriers vs IJN Carriers: The Pacific 1942. New York: Osprey. ISBN 978-1-84603-248-6;

[41]-Stüwe, Botho., p. 258 Das Ortungsignal der Me 163 B war relativ schwach ... difficult Radar target, absence of dihedral reflector (tailless). Peenemünde West (in German). Augsburg, Germany: Bechtermünz Verlag, 1999. ISBN 3-8289-0294-4;

[42]-Weiner, Tim (1997-08-23). "The $2 Billion Stealth Bomber Can't Go Out in the Rain". The New York Times. Retrieved 2007-12-18;

[43]-Willmott, H.P. (1983). The Barrier and the Javelin: Japanese and Allied Strategies, February to June 1942. United States Naval Institute Press. p. 616. ISBN 1-59114-949-5;

[44]-Wise, John (1850), A System of Aeronautics, Comprehending its Earliest Investigations, and Modern Practice and Art, Philadelphia, PA: Joseph A Speel, retrieved 2010-09-13;